Advances in Intelligent Systems and Computing

Volume 1004

The series "Advances in Intelligent Systems and Computing" contains publications on theory, applications, and design methods of Intelligent Systems and Intelligent Computing. Virtually all disciplines such as engineering, natural sciences, computer and information science, ICT, economics, business, e-commerce, environment, healthcare, life science are covered. The list of topics spans all the areas of modern intelligent systems and computing such as: computational intelligence, soft computing including neural networks, fuzzy systems, evolutionary computing and the fusion of these paradigms, social intelligence, ambient intelligence, computational neuroscience, artificial life, virtual worlds and society, cognitive science and systems, Perception and Vision, DNA and immune based systems, self-organizing and adaptive systems, e-Learning and teaching, human-centered and human-centric computing, recommender systems, intelligent control, robotics and mechatronics including human-machine teaming, knowledge-based paradigms, learning paradigms, machine ethics, intelligent data analysis, knowledge management, intelligent agents, intelligent decision making and support, intelligent network security, trust management, interactive entertainment, Web intelligence and multimedia.

The publications within "Advances in Intelligent Systems and Computing" are primarily proceedings of important conferences, symposia and congresses. They cover significant recent developments in the field, both of a foundational and applicable character. An important characteristic feature of the series is the short publication time and world-wide distribution. This permits a rapid and broad dissemination of research results.

** **Indexing: The books of this series are submitted to ISI Proceedings, EI-Compendex, DBLP, SCOPUS, Google Scholar and Springerlink** **

More information about this series at http://www.springer.com/series/11156

Enrique Herrera-Viedma ·
Zita Vale · Peter Nielsen ·
Angel Martin Del Rey ·
Roberto Casado Vara
Editors

Distributed Computing and Artificial Intelligence, 16th International Conference, Special Sessions

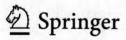
Springer

Editors
Enrique Herrera-Viedma
University of Granada
Granada, Spain

Zita Vale Ⓘ
Engineering Institute
Polytechnic of Porto
Porto, Portugal

Peter Nielsen
Department of Materials and Production,
The Faculty of Engineering and Science
Aalborg University
Aalborg, Denmark

Angel Martin Del Rey
Department of Applied Mathematics,
Faculty of Science
University of Salamanca
Salamanca, Salamanca, Spain

Roberto Casado Vara
BISITE Digital Innovation Hub
University of Salamanca
Salamanca, Spain

ISSN 2194-5357   ISSN 2194-5365   (electronic)
Advances in Intelligent Systems and Computing
ISBN 978-3-030-23945-9   ISBN 978-3-030-23946-6   (eBook)
https://doi.org/10.1007/978-3-030-23946-6

This Springer imprint is published by the registered company Springer Nature Switzerland AG
The registered company address is: Gewerbestrasse 11, 6330 Cham, Switzerland

Preface

The 16th International Conference on Distributed Computing and Artificial Intelligence 2019 is an annual forum that will bring together ideas, projects and lessons associated with distributed computing and artificial intelligence, and their application in different areas. Artificial intelligence is changing our society. Its application in distributed environments, such as the Internet, electronic commerce, environment monitoring, mobile communications, wireless devices, distributed computing, to mention only a few, is continuously increasing, becoming an element of high added value with social and economic potential, in industry, quality of life and research. These technologies are changing constantly as a result of the large research and technical effort being undertaken in both universities and businesses. The exchange of ideas between scientists and technicians from both the academic and industry sector is essential to facilitate the development of systems that can meet the ever-increasing demands of today's society.

The present edition brings together past experience, current work and promising future trends associated with distributed computing, artificial intelligence and their application in order to provide efficient solutions to real problems. This conference is a stimulating and productive forum where the scientific community can work towards future cooperation in distributed computing and artificial intelligence areas. Nowadays, it is continuing to grow and prosper in its role as one of the premier conferences devoted to the quickly changing landscape of distributed computing, artificial intelligence and the application of AI to distributed systems.

This year's technical program will present both high quality and diversity, with contributions in well-established and evolving areas of research. More than 120 papers were submitted to main and special sessions tracks from over 20 different countries (Algeria, Angola, Austria, Brazil, Colombia, France, Germany, India, Italy, Japan, the Netherlands, Oman, Poland, Portugal, South Korea, Spain, Thailand, Tunisia, UK and USA), representing a truly "wide area network" of research activity.

Moreover, DCAI'19 Special Sessions have been a very useful tool in order to complement the regular program with new or emerging topics of particular interest to the participating community. The DCAI'19 Special Sessions technical program has selected 17 papers and, as in past editions, it will be special issues in JCR-ranked journals such as Neurocomputing, and International Journal of Knowledge and Information Systems. Special Sessions that emphasize on multidisciplinary and transversal aspects, such as Advances on Demand Response and Renewable Energy Sources in Smart Grids (ADRESS), AI-driven methods for Multimodal Networks and Processes Modelling (AIMPM), Theoretical Foundations and Mathematical Models in Computer Science, Artificial Intelligence and Big Data (TMM-CSAIBD), have been especially encouraged and welcome. This volume also includes the 12 selected articles of the Doctoral Consortium.

This symposium is organized by the Osaka Institute of Technology and the University of Salamanca. The present edition was held in Avila, Spain, from 26 to 28 June 2019.

We thank the sponsors (IEEE Systems Man and Cybernetics Society Spain Section Chapter and the IEEE Spain Section (Technical Co-Sponsor), IBM, Indra, Viewnext, Global Exchange, AEPIA, APPIA and AIR institute) and the funding supporting of the Junta de Castilla y León, Spain, with the project *"Virtual-Ledgers-Tecnologías DLT/Blockchain y Cripto-IOT sobre organizaciones virtuales de agentes ligeros y su aplicación en la eficiencia en el transporte de última milla"* (Id. SA267P18-Project co-financed with FEDER funds), and finally, the Local Organization members and the Program Committee members for their hard work, which was essential for the success of DCAI'19.

Enrique Herrera-Viedma
Zita Vale
Peter Nielsen
Angel Martin Del Rey
Roberto Casado Vara

Organization

Honorary Chairmen

Masataka Inoue   President of Osaka Institute of Technology, Japan
Sigeru Omatu   Hiroshima University, Japan

Program Committee Chairs

Francisco Herrera   University of Granada, Spain
Kenji Matsui   Osaka Institute of Technology, Japan
Sara Rodríguez   University of Salamanca, Spain

Workshop Chair

Enrique Herrera Viedma   University of Granada, Spain

Doctoral Consortium Chair

Sara Rodríguez   University of Salamanca, Spain

Organizing Committee

Juan Manuel Corchado   University of Salamanca, Spain,
   Rodríguez   and AIR Institute, Spain
Sara Rodríguez González   University of Salamanca, Spain
Roberto Casado Vara   University of Salamanca, Spain
Fernando De la Prieta   University of Salamanca, Spain
Sonsoles Pérez Gómez   University of Salamanca, Spain
Benjamín Arias Pérez   University of Salamanca, Spain

Javier Prieto Tejedor   University of Salamanca, Spain,
                                 and AIR Institute, Spain
Pablo Chamoso Santos   University of Salamanca, Spain
Amin Shokri Gazafroudi   University of Salamanca, Spain
Alfonso González Briones   University of Salamanca, Spain,
                                 and AIR Institute, Spain
José Antonio Castellanos   University of Salamanca, Spain
Yeray Mezquita Martín   University of Salamanca, Spain
Enrique Goyenechea   University of Salamanca, Spain
Javier J. Martín Limorti   University of Salamanca, Spain
Alberto Rivas Camacho   University of Salamanca, Spain
Ines Sitton Candanedo   University of Salamanca, Spain
Daniel López Sánchez   University of Salamanca, Spain
Elena Hernández Nieves   University of Salamanca, Spain
Beatriz Bellido   University of Salamanca, Spain
María Alonso   University of Salamanca, Spain
Diego Valdeolmillos   University of Salamanca, Spain,
                                 and AIR Institute, Spain
Sergio Marquez   University of Salamanca, Spain
Guillermo Hernández   University of Salamanca, Spain
  González
Mehmet Ozturk   University of Salamanca, Spain
Luis Carlos Martínez   University of Salamanca, Spain,
  de Iturrate   and AIR Institute, Spain
Ricardo S. Alonso Rincón   University of Salamanca, Spain
Javier Parra   University of Salamanca, Spain
Niloufar Shoeibi   University of Salamanca, Spain
Zakieh Alizadeh-Sani   University of Salamanca, Spain
Jesús Ángel Román Gallego   University of Salamanca, Spain
Angélica González Arrieta   University of Salamanca, Spain
José Rafael García-Bermejo   University of Salamanca, Spain
  Giner
Pastora Vega Cruz   University of Salamanca, Spain
Mario Sutil   University of Salamanca, Spain
Ana Belén Gil González   University of Salamanca, Spain
Ana De Luis Reboredo   University of Salamanca, Spain

Contents

Special Session on Advances on Demand Response and Renewable Energy Sources in Smart Grids (ADRESS)

Special Session on Advances on Demand Response and Renewable Energy Sources in Smart Grids (ADRESS)

Smart Grid concepts are rapidly being transferred to the market and huge investments have already been made in renewable based electricity generation and in rolling out smart meters. However, the present state of the art does not ensure neither a good return of investment nor a sustainable and efficient power system. The work so far involves mainly larger stakeholders, namely power utilities and manufacturers and their main focus has been on the production and grid resources. This vision is missing a closer attention to the demand side and especially to the interaction between the demand side and the new methods for smart grid management.

Efficient power systems require, at all moments, the optimal use of the available resources to cope with demand requirements. Demand response programs framed by adequate business models will play a key-role in more efficient systems by increasing demand flexibility both on centralized and distributed models, particularly for the latter as renewable energy generation and storage is highly dependable of uncontrolled factors (such as wind and solar radiation) for which anticipated forecasts are subjected to significant errors.

The complexity and dynamic nature of these problems requires the application of advanced solutions to enable the achievement of relevant advancements in the state of the art. Artificial intelligence and distributed computing systems are, consequently, being increasingly embraced as a valuable solution. ADRESS aims at providing an advanced discussion forum on recent and innovative work in the fields of demand response and renewable energy sources integration in the power system. Special relevance is indorsed to solutions involving the application of artificial intelligence approaches, including agent-based systems, data-mining, machine learning methodologies, forecasting and optimization, especially in the scope of smart grids and electricity markets.

Organization

Organizing Committee

Zita Vale	Polytechnic of Porto, Portugal
Pedro Faria	Polytechnic of Porto, Portugal
Juan M. Corchado	University of Salamanca, Spain
Tiago Pinto	University of Salamanca, Spain
Pierluigi Siano	University of Salerno, Italy

Program Committee

Bo Norregaard Jorgensen	University of Southern Denmark, Denmark
Carlos Ramos	Polytechnic of Porto, Portugal
Cătălin Buiu	Politehnica University Bucharest, Romania
Cédric Clastres	Institut National Polytechnique de Grenoble, France
Dante I. Tapia	Nebusens, Spain
Frédéric Wurtz	Institut National Polytechnique de Grenoble, France
Georg Lettner	Vienna University of Technology, Austria
Germano Lambert-Torres	Dinkart Systems, Brazil
Gustavo Figueroa	Instituto de Investigaciones Eléctricas, Mexico
Ines Hauer	Otto-von-Guericke-University Magdeburg, Germany
Isabel Praça	Polytechnic of Porto, Portugal
István Erlich	University of Duisburg-Essen, Germany
Jan Segerstam	Empower IM Oy, Finland
José Rueda	Delft University of Technology, The Netherlands
Juan Corchado	University of Salamanca, Spain
Juan F. De Paz	University of Salamanca, Spain
Kumar Venayagamoorthy	Clemson University, USA
Lamya Belhaj	l'Institut Catholique d'Arts et Métiers, France
Nikolaus Starzacher	Discovergy, Germany
Nikos Hatziargyriou	National Technical University of Athens, Greece
Marko Delimar	University of Zagreb, Croatia
Nouredine Hadj-Said	Institut National Polytechnique de Grenoble, France
Pablo Ibarguengoytia	Instituto de Investigaciones Eléctricas, Mexico
Paolo Bertoldi	European Commission, Institute for Energy and Transport, Belgium
Pierluigi Siano	University of Salerno, Italy
Pedro Faria	Polytechnic of Porto, Portugal
Peter Kadar	Budapest University of Technology and Economics, Hungary
Pierre Pinson	Technical University of Denmark, Denmark
Rodrigo Ferreira	Intelligent Sensing Anywhere, Portugal
Stephen McArthur	University of Strathclyde, Scotland, UK

Tiago Pinto   Polytechnic of Porto, Portugal
Xavier Guillaud   École Centrale de Lille, France
Zbigniew Antoni Styczynski   Otto-von-Guericke-University Magdeburg,
                                         Germany
Zita Vale   Polytechnic of Porto, Portugal

Energy Consumption Forecasting Using Ensemble Learning Algorithms

Jose Silva[1], Isabel Praça[1], Tiago Pinto[1(✉)], and Zita Vale[2]

[1] GECAD Research Group, Institute of Engineering,
Polytechnic of Porto (ISEP/IPP), Porto, Portugal
{icp, tcp}@isep.ipp.pt
[2] Polytechnic of Porto (ISEP/IPP), Porto, Portugal
zav@isep.ipp.pt

Abstract. The increase of renewable energy sources of intermittent nature has brought several new challenges for power and energy systems. In order to deal with the variability from the generation side, there is the need to balance it by managing consumption appropriately. Forecasting energy consumption becomes, therefore, more relevant than ever. This paper presents and compares three different ensemble learning methods, namely random forests, gradient boosted regression trees and Adaboost. Hour-ahead electricity load forecasts are presented for the building N of GECAD at ISEP campus. The performance of the forecasting models is assessed, and results show that the Adaboost model is superior to the other considered models for the one-hour ahead forecasts. The results of this study compared to previous works indicates that ensemble learning methods are a viable choice for short-term load forecast.

Keywords: Electricity consumption · Short-term load forecast ·
Ensemble learning methods · Forecasting

1 Introduction

Electricity demand forecasting is an important task for the agents and companies involved in the electricity market. The features of the electricity, that is a non-storable product, and also the rules of this competitive market create the need of accurate predictions of electricity demand in order to anticipate decisions [1]. Thus, Electric Power Load Forecasting (EPLF) is a crucial process in the planning of electricity industry and the operation of electric power systems. The EPLF is classified in terms of the planning horizon's duration up: 1 day for short-term load forecasting (STLF), 1 day to 1 year for medium-term load forecasting (MTLF), and 1–10 years for long-term load forecasting (LTLF) [2]. A vast number of studies have developed accurate models in recent years. There are usually two approaches. The traditional statistical approach like multiple linear regression (MLR) used in [3] that achieved a 3.99% Mean Absolute Percentage Error (MAPE) on hourly electric load forecast. However, traditional statistical methods often result in lower accuracy because they are inadequate to fully model the complex nature of electricity demand. Artificial Intelligence (AI) techniques are more reliable due to their ability to identify non-linear relationships between

© Springer Nature Switzerland AG 2020
E. Herrera-Viedma et al. (Eds.): DCAI 2019, AISC 1004, pp. 5–13, 2020.
https://doi.org/10.1007/978-3-030-23946-6_1

dependent and independent variables. Artificial Neural Networks (ANN) [4], Support Vector Machines (SVM) [5], Random Forest [6] and Stochastic Gradient Boosting [7] are popular AI techniques for STLF.

This paper presents and compares the application of three ensemble learning methodologies to the problem of STLF. The applied algorithms are Random Forests (RF), Gradient Boosted Regression Trees (GBR) and Adaboost (AR2). These algorithms are adapted and applied to the forecasting of energy consumption of an office building. The objective of this study is to forecast a better profile of the energy consumption for the coming hours. Results from the proposed approach are compared to those achieved in previous studies, using different techniques, namely three fuzzy based systems: an Hybrid Neural Fuzzy Inference System (HyFIS) [8], the Wang and Mendel's Fuzzy Rule Learning Method (WM) [9]; and SVM [10]. The case study is based on real data referring to the electricity consumption of a campus building of ISEP/GECAD - Research Group on Intelligent Engineering and Computing for Advanced Innovation and Development.

After this introductory section, Sect. 2 presents the formulation and explanation of the proposed approach, Sect. 3 presents the achieved results and discusses their comparison to the results achieved by previous methods. Section 4 presents the most relevant conclusions and contributions of this work

2 Materials and Methods

This paper presents and discusses the implementation of three ensemble-learning methods to forecast the electricity consumption of an office building. The electricity consumption form building N of the GECAD research center located in ISEP/IPP, Porto, Portugal is used in this work. The ensemble approach has been developed based on Python programming language. The implementation details and results of this work are discussed and compared in the following sections.

Ensemble methods combine the predictions of several base estimators built with a given learning algorithm in order to improve generality and robustness over a single estimator. They can be distinguished in two categories; averaging methods, where the main idea is to build several estimators independently and then to average their predictions, and boosting methods where the driving principle is to combine several weak models to produce a powerful ensemble. Examples of the first category are the bagging methods and RF while the second category includes methods like Adaboost and GBRT.

2.1 Random Forests (RF)

Random forests is an ensemble learning method for classification and regression. In RF each tree in ensemble is generated by randomly selecting the attributes to split at each node and these features on training set are used to estimate best split. As a result of this randomness, the bias of the forest usually slightly increases (with respect to the bias of a single non-random tree) but, due to averaging, its variance also decreases, often more than compensating for the increase in bias, resulting an overall better model. Let us assume that, the training data contain K instances and M set of features. We are given n, where n is the set of variable $n \in N$. Learning set is created by choosing k instances from K with replacement, and the remaining instances are used to estimate the error of the model. At each node, randomly select n variables and make split decisions based on these variables. At the end fully grown unpruned tree is created. In this study we used the implementation of Python's scikit-learn library [11], which combines classifiers by averaging their probabilistic prediction, instead of letting each classifier vote for a single class. For our forecasting task we built a RF regressor of 200 trees with nodes which were expanded until all leaves were pure. The minimum number of samples required to split an internal node has been set to two.

2.2 Gradient Boosted Regression Trees (GBR)

GBRT is a machine learning technique that generates a prediction model in the form of an ensemble of weak prediction models which are typically decision trees. It builds the model sequentially and generalizes them by allowing optimization of an arbitrary differentiable loss function. It is mainly a regression technique invented by Jerome H. Friedman in 1999 [12]. GBR takes into account additive models of the form (1):

$$F(x) = \sum_{m=1}^{M} \gamma\, mhm(x) \tag{1}$$

where $hm(x)$ are the principle functions, which are called weak learners in the context of boosting and γm the step length that is chosen using line search (2):

$$\gamma m = argmin_y \sum_{i=1}^{n} L\left(yi, F_{m-1}(xi) - y\frac{\theta L(yi, Fm - 1(xi))}{\theta Fm - 1(xi)} \right) \tag{2}$$

Similarly to other boosting algorithms GBRT builds the additive model in a forward stage wise approach (3):

$$Fm(x) = Fm - 1(x) + \gamma mhm(x) \tag{3}$$

At each stage the decision tree $hm(x)$ is chosen to minimize the loss function L given the current model $Fm - 1$ and its fit $Fm - 1(xi)$, as in (4).

$$Fm(x) = Fm - 1(x) + argmin_h \sum_{i=1}^{n} L(yi, Fm - 1(xi) - h(x)) \tag{4}$$

The basic idea for solving this minimization problem is to use the steepest descent which is the negative gradient of the loss function evaluated at the current model $Fm - 1$. This can be estimated as in (5):

$$Fm(x) = Fm - 1(x) + \gamma m \sum_{i=1}^{n} \nabla_F L(yi, Fm - 1(xi)) \qquad (5)$$

To forecast our target values, we built a GBRT model with 1400 boosting stages using Python's scikit-learn package. Since gradient boosting is fairly robust to over-fitting, a large number of stages usually performs better. Scikit-learn supports many different loss function. We experienced all of them and found that the Least absolute deviations (LAD) function works better for our model.

To improve the performance of our model we also tuned the maximum depth of the individual regression estimators. This depth limits the number of nodes in the tree. Maximum depth of size 10 produced the best results. In addition, we set the minimum number of samples required to split an internal node to 2 and the learning rate to 0.2. The learning rate scales the step length of the gradient decent procedure. The parameter v in the following equation is the learning rate:

$$Fm(x) = Fm - 1(x) + v\gamma m hm(x) \qquad (6)$$

The above regularization technique was proposed in [13] and scales the contribution of each weak learner by v.

2.3   AdaBoost

Another used ensemble estimator is AdaBoost.R2 [14] which is a modified regression version of the famous AdaBoost ensemble estimator [15]. It sequentially fits estimators and each subsequent estimator concentrates on the samples that were predicted with higher loss.

The used algorithm implemented in [11] slightly differs from [14] as it allows to use the weights directly in the fitted estimator and not only for weighted sampling of features, as follows:

1. start algorithm t = 0
2. To each training sample assign initial weight (7)

$$w_i^t := 1, i = 1, 2, \ldots, m \qquad (7)$$

3. fit estimator t to the weighted training set with weights w_i^t
4. compute prediction $\widehat{y}_i^t$ using the estimator t for each sample i
5. compute loss li for each training sample (8)

$$l_i^t = loss(|\widehat{y}_i^t - yi|) \qquad (8)$$

6. calculate average loss $\overline{l^t}$

7. calculate confidence β^t for the estimator (low β^t means high confidence in estimator t)
8. update weights of training samples (9)

$$w_i^{t+1} = w_i^t \cdot (\beta^t)^{(1-l_i^t)}, i = 1, 2, \ldots, m \tag{9}$$

9. $t = t + 1$ continue to step 3 while the average loss $\bar{l}^t < 0.5$.

To forecast our target values, we built an AR2 model with 1400 boosting stages using Python's scikit-learn package and the learning rate has been set to 0.01. We experienced that square loss function works better for our model [14] (10):

$$l_i^t = \frac{\left|\widehat{y}_i^t - yi\right|^2}{D^2} \tag{10}$$

where D is defined as (11):

$$D = sup\{\left|\widehat{y}_i^t - yi\right|, , i \in \{1, 2, \ldots m\}\} \tag{11}$$

3 Case Study

3.1 Dataset Description

This study uses real data from ISEP/GECAD building N monitoring, an office building with a daily usage from around 30 researchers. The data is available in a SQL server that stores the information given by five energy analyzers, each of them stores the electricity consumption data coming from sockets, lighting and HAVAC from a part of the building, in a 10 s time interval. A java base application has been developed in this implementation. Which collect the data from the SQL server and calculate the average of the total electricity consumption of the building N – ISEP/GECAD per each hour. This application also creates a new .csv file in a format that can be used as the input of the forecast method.

The ensemble methods have been implemented for this forecast with various strategies presented in Table 2. The various strategies combine the features extracted from the 10 days before the hour which is meant to be forecasted presented in A total of 15 features has been generated. Table 1 presents the features used for training and testing of the building N consumption dataset.

Table 1. The results of this method are shown and compared to those of previous works.

3.2 Results

In order to test these methods, the electricity consumption from 00:00 until 23:00 of the date 5/4/2018 is forecasted. MAPE is used for error calculation as means to compare the forecasted values and the real values of each hour.

Table 1. Generated features

Feature description	Nomenclature
Consumption of the 3 previous hours	$Z_{t-1} \, to \, Z_{t-3}$
Hour of the day	H_t
Month of the year	M_t
Day of the month	D_t
Year	Y_t
Day of the week	DoW_t
Environmental temperature of the hour (°C)	T_t
Environmental temperature from the previous 3 h	$T_{t-1} to T_{t-3}$
Consumption at the same hour from the 2 previous weeks	$Z_{t-168} \, and \, Z_{t-336}$
Environmental Humidity of the hour	Hu_t

A total of 15 features has been generated. Table 1 presents the features used for training and testing of the building N consumption dataset.

The combination of the generated features in Table 1 resulted in the creation of seven training strategies, which are presented in Table 2.

Table 2. Training strategies

Strategy #	$Z \sim *$ (Consumption over …)
1	$Z \sim D_t * M_t * H_t * Y_t * DoW_t$
2	$Z \sim D_t * M_t * H_t * Y_t * DoW_t * T_t$
3	$Z \sim D_t * M_t * H_t * Y_t * DoW_t * T_t * Z_{t-1} \, to \, Z_{t-3}$
4	$Z \sim D_t * M_t * H_t * Y_t * DoW_t * T_t * Z_{t-1} \, to \, Z_{t-3} * T_{t-1} \, to \, T_{t-3}$
5	$Z \sim D_t * M_t * H_t * Y_t * DoW_t * T_t * Z_{t-168} \, and \, Z_{t-336}$
6	$Z \sim D_t * M_t * H_t * Y_t * DoW_t * T_t * Z_{t-1} \, to \, Z_{t-3} * T_{t-1} \, to \, T_{t-3} * Z_{t-168} \, and \, Z_{t-336}$
7	$Z \sim D_t * M_t * H_t * Y_t * DoW_t * T_t * Hu_t$

In Table 2 the strategies were designed to test the performance of different combinations of variables. The MAPE error results for each ensemble method and each strategy are presented in Fig. 1.

As Fig. 1 shows, the forecasting error of the third training strategy presents the best performance in all of the ensemble methods. It is noticeable that the use of the environmental temperature is better in all the scenarios but the lagged features of the environmental temperature produces a higher error than not using it. In addition, the use of lagged features of consumption performs better than not using it, the same happens to the environmental humidity.

The study presented in [16] addresses the electricity consumption forecast based on fuzzy rules methods of the same location as this study, namely using HyFIS, WM and SVM. In order to compare the results of the ensemble methods with the ones addressed in [16] we trained our models using the third training strategy and forecasted the 24 h of 10/4/2018. The comparison between the methods is shown in Fig. 2 and Table 3.

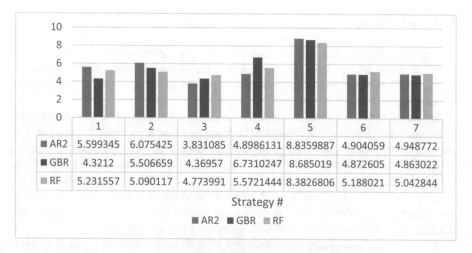

Strategy #	1	2	3	4	5	6	7
■ AR2	5.599345	6.075425	3.831085	4.8986131	8.8359887	4.904059	4.948772
■ GBR	4.3212	5.506659	4.36957	6.7310247	8.685019	4.872605	4.863022
■ RF	5.231557	5.090117	4.773991	5.5721444	8.3826806	5.188021	5.042844

■ AR2 ■ GBR ■ RF

Fig. 1. Average forecasting errors of the AR2, GBR and RF

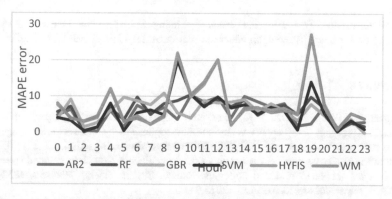

Fig. 2. Comparison between the results of AR2, RF and GBR and the results in [16]

Table 3. Average forecasting methods errors

	AR2	RF	GBR	SVM	HyFIS	WM
MAPE %	5,34	6,11	6,07	5,82	7,88	7,92

The comparison between the fuzzy rule based methods and the ensemble methods in Fig. 2 and Table 3 shows that the AR2 method has the lower average forecasting error between these methods and can achieve better forecasting results than the other methods.

4 Conclusions

This paper addresses the use of ensemble learning methodologies to forecast the electricity consumption of an office building in the following hours. This method uses the value of the electricity consumption from previous days to preview this value for the next hour.

By comparing the results of the AR2 method to the results of the SVM and some fuzzy rule-based methods, it is possible to conclude that it provides a more accurate consumption forecast. Additionally, the results presented in this paper in comparison to the results of the fuzzy rule-based methods presented in [16], namely SVM, HyFIS and WM, tend to forecast more reliable values and all the calculated errors are closer to the average error with exception to SVM that outperforms RF and GBR. The use of environmental variables such as a humidity and temperature proved to be useful in this experiment as well as the use of lagged features of previous hours.

As future work, we suggest the inclusion of additional exogenous variables in our models, such as direct solar irradiation or thermal sensation in order to improve the forecast accuracy.

Acknowledgements. This work has received funding from National Funds through FCT (Fundaçao da Ciencia e Tecnologia) under the project SPET – 29165, call SAICT 2017.

References

1. Zhang, X., Wang, J., Zhang, K.: Short-term electric load forecasting based on singular spectrum analysis and support vector machine optimized by Cuckoo search algorithm. Electr. Power Syst. Res. **146**, 270–285 (2017). https://doi.org/10.1016/j.epsr.2017.01.035
2. Raza, M.Q., Khosravi, A.: A review on artificial intelligence based load demand forecasting techniques for smart grid and buildings. Renew. Sustain. Energy Rev. **50**, 1352–1372 (2015). https://doi.org/10.1016/j.rser.2015.04.065
3. Saber, A.Y., Alam, A.K.M.R.: Short term load forecasting using multiple linear regression for big data. In: 2017 IEEE Symposium Series on Computational Intelligence (SSCI), pp. 1–6 (2017)
4. Pinto, T., Sousa, T.M., Vale, Z.: Dynamic artificial neural network for electricity market prices forecast. In: 2012 IEEE 16th International Conference on Intelligent Engineering Systems (INES), pp. 311–316 (2012)
5. Pinto, T., Sousa, T.M., Praça, I., et al.: Support Vector Machines for decision support in electricity markets' strategic bidding. Neurocomputing **172**, 438–445 (2016). https://doi.org/10.1016/j.neucom.2015.03.102
6. Ahmad, T., Chen, H.: Nonlinear autoregressive and random forest approaches to forecasting electricity load for utility energy management systems. Sustain Cities Soc. **45**, 460–473 (2019). https://doi.org/10.1016/j.scs.2018.12.013
7. Touzani, S., Granderson, J., Fernandes, S.: Gradient boosting machine for modeling the energy consumption of commercial buildings. Energy Build **158**, 1533–1543 (2018). https://doi.org/10.1016/j.enbuild.2017.11.039

8. Osório, G.J., Matias, J.C.O., Catalão, J.P.S.: Short-term wind power forecasting using adaptive neuro-fuzzy inference system combined with evolutionary particle swarm optimization, wavelet transform and mutual information. Renew. Energy **75**, 301–307 (2015). https://doi.org/10.1016/j.renene.2014.09.058
9. Gou, J., Hou, F., Chen, W., et al.: Improving Wang–Mendel method performance in fuzzy rules generation using the fuzzy C-means clustering algorithm. Neurocomputing **151**, 1293–1304 (2015). https://doi.org/10.1016/j.neucom.2014.10.077
10. Du, P., Wang, J., Yang, W., Niu, T.: Multi-step ahead forecasting in electrical power system using a hybrid forecasting system. Renew Energy **122**, 533–550 (2018). https://doi.org/10.1016/j.renene.2018.01.113
11. Pedregosa, F., Varoquaux, G., Gramfort, A., et al.: Scikit-learn: machine learning in python. J. Mach. Learn. Res. **12**, 2825–2830 (2011)
12. Friedman, J.H.: Stochastic gradient boosting. Comput. Stat. Data Anal. **38**, 367–378 (2002). https://doi.org/10.1016/S0167-9473(01)00065-2
13. Friedman, J.H.: Greedy function approximation: a gradient boosting machine. Ann. Stat. **29**, 1189–1232 (2001)
14. Drucker, H.: Improving regressors using boosting techniques. In: Proceedings of the Fourteenth International Conference on Machine Learning, pp. 107–115. Morgan Kaufmann Publishers Inc., San Francisco (1997)
15. Freund, Y., Schapire, R.E.: A decision-theoretic generalization of on-line learning and an application to boosting. J. Comput. Syst. Sci. **55**, 119–139 (1997). https://doi.org/10.1006/jcss.1997.1504
16. Jozi, A., Pinto, T., Praça, I., Vale, Z.: Day-ahead forecasting approach for energy consumption of an office building using support vector machines. In: 2018 IEEE Symposium Series on Computational Intelligence (SSCI), pp. 1620–1625 (2018)

Study of Multi-Tariff Influence
on the Distributed Generation Remuneration

Cátia Silva$^{(\boxtimes)}$, Pedro Faria, and Zita Vale

GECAD - Research Group on Intelligent Engineering and Computing
for Advanced Innovation and Development, IPP - Polytechnic Institute of Porto,
Rua DR. Antonio Bernardino de Almeida, 431, 4200-072 Porto, Portugal
{cvcds,pnf,zav}@isep.ipp.pt

Abstract. The energy market, with the introduction of the smart grids concept, opens the door to small distributed energy resources. However, these resources introduce an added level of difficulty to market management, requiring an entity to aggregate and manage them optimally. This paper proposes an approach that integrates these small resources. The methodology is composed of optimal scheduling, aggregation and remuneration based on aggregation. The method chosen for aggregation is k-means. In relation to previous works, the innovation goes through the multi-period and the comparison that this can have in the formation of groups. Thus, three scenarios were created: Whole Week, Work Days and Weekend. Profiles were added for 548 units of DG. The justification for the formation of groups will be a fairer remuneration and according to the contribution of each resource to the management of the network.

Keywords: Aggregation · Clustering · Distributed generation

1 Introduction

Bidirectional communication will improve the electric distribution network and add great value to the future energy market [1]. The concept of Smart Grids will introduce new opportunities for small-scale resources, so far without any information about the transactions that existed in the network. Consumers will be able to enter De-mand Response (DR) programs where they can respond to real-time signals to change their consumption [2]. There is also an incentive to use distributed generation (DG) units, not only because of their environmental benefits but also because can help reduce transport losses. Although, this resources will introduce uncertainty to the management of the network being necessary an entity to manage them in a proper way - Virtual Power Player (VPP) [3]. The need for tools and models that are able to handle this type and amount of information will be needed. Several works use artificial intelligence in this type of models [4] and the authors propose a management methodology to assist VPP in their task. This paper is the development of previous work [5]. In the first place,

The present work was done and funded in the scope of the following projects: COLORS Project PTDC/EEI-EEE/28967/2017 and UID/EEA/00760/2019 funded by FEDER Funds through COMPETE program and by National Funds through FCT.

© Springer Nature Switzerland AG 2020
E. Herrera-Viedma et al. (Eds.): DCAI 2019, AISC 1004, pp. 14–19, 2020.
https://doi.org/10.1007/978-3-030-23946-6_2

optimal scheduling of resources is proposed, and after an aggregation of resources through a clustering method and final remuneration through a group tariff. In this paper, the aggregation will be done for several time periods and the goal is to understand, testing different numbers of groups, which is the most beneficial situation for VPP. The remuneration made in this way ensures that all resources are remunerated fairly. In this way it is possible to motivate the continued collaboration of resources with VPP.

Section 1 presents a brief introduction to the theme developed throughout the Paper. The second section details the proposed methodology. Section 3 presents the case study and Sect. 4 the results obtained. Finally, Sect. 5 presents the conclusions obtained.

2 Methodology

As already mentioned, the objective of this methodology is to optimally manage all the small resources associated with the entity that aggregate them. This approach will have several advantages: VPP will have enough energy to enter the market and these resources, which until now had no direct contact with the energy market, could have. Figure 1 presents a scheme of the methodology proposed by the authors.

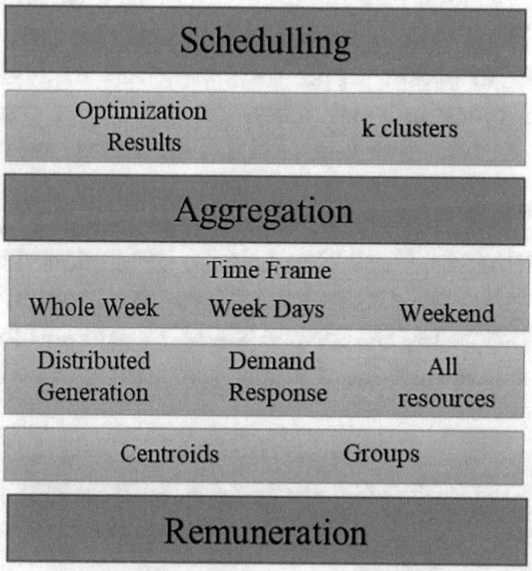

Schedulling		
Optimization Results		k clusters
Aggregation		
	Time Frame	
Whole Week	Week Days	Weekend
Distributed Generation	Demand Response	All resources
Centroids		Groups
Remuneration		

Fig. 1. Proposed methodology

The first phase, Scheduling, is an optimization problem in which the main objective is to minimize operating costs. The resources that may be associated with the VPP are consumers belonging to DR programs - both incentive based (IDR) and real time pricing (RTP), DG units and suppliers. In the event that DG resources do not meet consumers' needs, suppliers will be used. This optimization is subject to price and operating restrictions as well as operational restrictions that are imposed by the VPP to achieve its objectives.

The aggregation phase is in focus in this paper since the objective of this study is to aggregate in three different temporal situations the resources - Whole Week (WW), Work Days (WD) and Weekend (W) and realize which will be the most beneficial. Thus, the input data of this phase are the results of the previous optimization and the number of groups that the VPP intends to see formed - k. The proposal will be the use of a clustering method to accomplish this task. The method selected was k-means. This algorithm defines that at each iteration the distance between the points in the database and the center of each group is calculated, with the said point being assigned to the group with the lowest distance. The aggregation is done separately in this paper, that is, each resource is aggregated by type - DR or DG.

Finally, the remuneration phase is essential in this method. The groups formed served to provide a fair remuneration for the resources and to encourage them to continue to collaborate with the VPP. The authors proposed the remuneration through the maximum tariff for each group. For example, a resource that has a low rate will receive the same way as the rest of the group.

3   Case Study

The proposed methodology will be studied through the case study presented in this section. The database belongs to a distribution network of 30 kV, with a maximum capacity of 90MVA. The VPP is responsible for specifying different characteristics for each resource, in order to obtain proper results for this methodology. There are five types of consumers: Domestic, Small Commerce, Medium Commerce, Large Commerce and Industrial. In total there are 20310 consumers and a total capacity of 21 354.36 kWh.

As for DG units, there are seven different ones: Small Hydro, Waste-to-energy, Wind, Photovoltaic, Biomass, Fuell Cell and Co-generation. The number of units is 548 and the total capacity is 25 388.79 kWh. Consumer profiles, such as DG units and Consumers from DR programs, are for periods of 15 min and for one week - January 2–8, 2018. Figure 2. shows how the different scenarios will be studied.

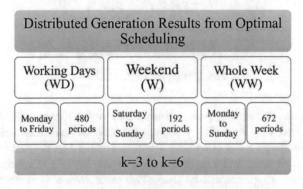

Fig. 2. Formation of case study

4   Results

This section presents the results for the second phase of the methodology. This phase was performed using software R and using the k-means as clustering method. Due to the space limitation, it will not be possible to exhaustively analyze all k chosen or all resources. Thus, only k = 6 will be developed in more detail. The resources chosen for this study were the DG units.

The k-means method used had as output: the group associated with each resource analyzed and the centroid of the group. Figure 3 presents the centroids for each group found. The centroid value allows us to estimate the mean value of the reduced power

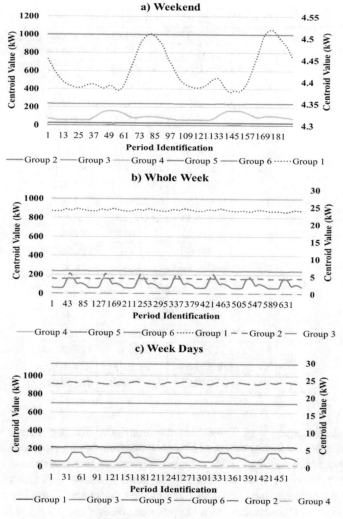

Fig. 3. Centroid value per group

by the elements of a given group. This value may be very useful for VPP in the case of adding new resources to existing groups.

The figures have two y-axis in order to show all curves, due to the difference in scale. Thus, to differentiate the curves from each other, the dashed lines are from the second axis. After analyzing the figure, it is possible to conclude that for all scenarios there is the possibility of dividing the groups into three different cases: there are groups with small reductions (below 30 kW), medium reductions (between 60 and 250 kW) and the largest reductions (1000 kW). Table 1 shows the compensation figures for groups. These values were calculated through remuneration with the maximum tariff of a given group by the contribution of each resource of DG belonging to the group.

Table 1. Remuneration per group and total

	Group	WW (m.u.)	WD (m.u.)	W (m.u.)
k = 3	1	1 276 923.97	915 049.45	93 600.00
	2	218 354.66	156 038.80	361 874.52
	3	327 600.00	234 000.00	62 315.86
	Total	1 822 878.63	1 305 088.25	517 790.38
k = 4	1	26 524.56	19 089.73	62 315.86
	2	427 620.15	234 000.00	93 600.00
	3	327 600.00	306 406.44	121 213.71
	4	218 354.66	156 038.80	7 434.84
	Total	1 000 099.37	715 534.97	284 564.40
k = 5	1	26 524.56	138 259.33	93 600.00
	2	327 600.00	98 280.00	35 896.10
	3	100 346.54	19 089.73	7 434.84
	4	125 879.33	300 710.97	28 670.44
	5	419 643.99	159 120.00	118 933.02
	Total	999 994.42	715 460.03	284 534.40
k = 6	1	414 532.96	132 638.80	2 449.95
	2	8 823.16	296 881.62	69.69
	3	0.00	98 280.00	62 315.86
	4	327 600.00	19 089.73	3 465.02
	5	12 729.55	9 264.53	117 651.34
	6	218 354.66	159 120.00	93 600.00
	Total	982 040.32	715 274.68	279 551.86

Through the analysis of Table 1, it was concluded that the case k = 6 for all scenarios obtains the lowest values of total remuneration. In order to prove the viability of the proposed methodology, different types of remuneration were test and compared. The Table 2 shows the final remuneration for other methods. In this way, method 1 represents the remuneration in a individual way, method 2 uses a formula present by [6] and method 3 the remuneration with the proposed methodology for k = 6.

Table 2. Remuneration per group and total

Method	WW (m.u.)	WD (m.u.)	W (m.u.)
1	978 404.56	698 482.79	279 921.77
2	685 879.51	489 917.36	195 962.15
3	982 040.32	715 274.68	279 551.86

After analyzing the Table 2, the method 2 got the lowest value of remuneration. This formula depends has a parameter that depends of the resource. In this way, depending on fluctuation of this value, the incentive can change. The first and third method have a final remuneration value more similar but, the last one has the higher value. Since the main purpose of this phase is to reward fairly and still motivate the resource to the continuous, the higher the incentive is, the greater the possibility of the participation of the resource. In this way, regardless the disposition of the resource to contribute, it will always receive according to what has been reduced, through the maximum tariff of the group in which it is inserted.

5 Conclusions

This paper presents an approach that assists VPP in the task of optimally managing small resources, such as DG units and DR consumers. The focus was on aggregating resources using the clustering algorithm functionalities in the formation of groups in which their elements would have similar characteristics. This would be useful for the remuneration of the resources associated with the VPP, since, by remunerating them by trained groups, they would receive in a fairer way because a specific tariff would be created. Previous work covered this work for only one period. The innovation of this paper is to introduce the concept of multiperiod to perceive the influence of this in the creation of the groups and in the final remuneration of the resources.

References

1. Yu, N., Wei, T., Zhu, Q.: From passive demand response to proactive demand participation. IEEE Int. Conf. Autom. Sci. Eng. **2015**, 1300–1306 (2015)
2. Lujano-Rojas, J.M., Monteiro, C., Dufo-López, R., Bernal-Agustín, J.L.: Optimum residential load management strategy for real time pricing (RTP) demand response programs. Energy Policy **45**, 671–679 (2012)
3. Faria, P., Spínola, J., Vale, Z.: Aggregation and remuneration of electricity consumers and producers for the definition of demand-response programs. IEEE Trans. Ind. Informatics **12**(3), 952–961 (2016)
4. Chaouachi, A., Kamel, R.M., Andoulsi, R., Nagasaka, K.: Multiobjective intelligent energy management for a microgrid. IEEE Trans. Ind. Electron. **60**(4), 1688–1699 (2013)
5. Silva, C., Faria, P., Vale, Z.: Clustering support for an aggregator in a smart grid context. In: 18th International Conference on Hybrid Intelligent Systems (HIS) (2018)
6. Cabrera, N.G., Gutierrez-Alcaraz, G.: Evaluating demand response programs based on demand management contracts. In: IEEE Power and Energy Society General Meeting, pp. 1–6 (2012)

Lighting Consumption Optimization in a SCADA Model of Office Building Considering User Comfort Level

Mahsa Khorram, Pedro Faria$^{(\boxtimes)}$, and Zita Vale

GECAD – Research Group on Intelligent Engineering and Computing
for Advanced Innovation and Development, Institute of Engineering,
Polytechnic of Porto (ISEP/IPP), Porto, Portugal
{makgh, pnfar, zav}@isep.ipp.pt

Abstract. Due to the high penetration of the buildings in energy consumption, the use of optimization algorithms plays a key role. Therefore, all the producers and prosumers should be equipped with the automation infrastructures as well as intelligent decision algorithms, in order to perform the management programs, like demand response. This paper proposes a multi-period optimization algorithm implemented in a multi-agent Supervisory Control and Data Acquisition system of an office building. The algorithm optimizes the lighting power consumption of the building considering the user comfort constraints. A case study is implemented in order to validate and survey the performance of the implemented optimization algorithm using real consumption data of the building. The outcomes of the case study show the great impact of the user comfort constraints in the optimization level by respect to the office user's preferences.

Keywords: Multi-period optimization · SCADA · User comfort ·
Office building

1 Introduction

Every day, a lot of energy is lost by the negligence of people around the world. Sometimes the unimportant actions during a day can be the terminator of the environment and earth at the end [1]. That is why the world is moving towards comprehensive automation and smart infrastructure in the buildings, in order to prevent the loss of energy as much as possible [2]. In addition to these facilities, Demand Response (DR) programs organize the user's consumption pattern as a generic and systematic program according to electricity price variations or technical issues with considering consumers and producers interests. DR programs have a desirable variety which is divided into two main groups, namely price-based demand response and incentive-based demand response [3].

The present work was done and funded in the scope of the following projects: COLORS Project PTDC/EEI-EEE/28967/2017 and UID/EEA/00760/2019 funded by FEDER Funds through COMPETE program and by National Funds through FCT.

The buildings are responsible for 40% of world energy consumption which is increasing every day [4]. Among all types of buildings, office buildings can be considered as a more flexible option for implementing DR programs, since usually they have significant energy consumption, and also in some cases can be more equipped to automation infrastructure than residential houses.

Recently, the main concern in energy minimization topics is respect to user comfort while energy consumption is optimizing [5]. keeping a balance between energy minimization and user preferences need a formulation with precise restrictions in order to observe optimization purposes and user easement at the same time [6].

Mostly in office buildings, more attention is paid to Air Conditioners (AC) while 29% of total energy consumption in office buildings belongs to the lighting system [7]. The lights of an office building can be considered as flexible loads for reduction and curtailment if they are fully controllable and reducible by existing equipment. Supervisory Control And Data Acquisition (SCADA) system play a key role in DR implementation since it offers various advantages in order to have automatic load control in different types of buildings [8]. For instance, the SCADA system can dominate the lights of the illumination system which they are fully controllable via the Digital Addressable Lighting Interface (DALI) [8].

In this matter, SCADA systems can be integrated with the Multi-Agent Systems (MAS) for improving the overall system performance. If the SCADA system is equipped with the MAS, various types of optimization algorithms could be solved and utilized by the model in order to control and manage the resources controlled by the SCADA system [9]. Agent-based SCADA models provide more flexibility and adaptability [10].

This paper proposes a multi-period optimization algorithm for the lighting system of an office. The algorithm focuses on the minimization of the power consumption of the lights with respect to user comfort. The algorithm is implemented in an agent-based SCADA system installed in the office building. All the parameters in the building, such as the consumption of each light and total consumption as well, are monitored through this SCADA system.

Several studies have been done in the context of building energy optimization. In [7], the authors proposed a smart lighting control based on internal mode controller of an artificial neural network which tries to maintain occupant's preferences while are using natural light at the same time. In [8], presented a SCADA-based model focused on the lights and ACs consumption of the building for participating in DR events. In [11], ACs and lights consumption is managed and minimized under Real Time Pricing (RTP) tariffs in a MAS based SCADA model. A lighting consumption optimization has been proposed in [12] by considering the renewable resources. In [13], the user satisfaction provided based on time and device, while user budget is considered. However, the focus of this paper is to study the impact of the user comfort constraints considered for the proposed optimization algorithm. The outcomes of optimization would be compared and surveyed with and without considering the user comfort constraints.

After this section, the optimization algorithm and the implemented methodology is explained in Sect. 2. A case study is demonstrated in Sect. 3 in order to validate the performance of the proposed optimization algorithm, and the gained results will be compared in the same section. Finally, Sect. 4 describes the main conclusions of the work.

2   Optimization Algorithm

This section presents the optimization algorithm implemented in the SCADA system. As it was described, the SCADA model is agent-based, with a various number of agents and players. The system was developed by the authors in the scope of their previous works [11], and this section focuses only on the Optimizer agent and the implemented optimization algorithm on this agent. More details and information regarding the SCADA model and the other agents are available on [11].

The main purpose of the algorithm is minimizing the power consumption of the lights with respect to user comfort. Each light participates in minimization call from optimizer agent as a component of the system. For each light, an importance weight is dedicated by numbers between 0 and 1, in order to define the priority of them. These numerical criteria are dependence to several conditions such as preferences of the user, the location of the light and operation of natural daylight. These priority numbers observe the user comfort to some extent but, more restrictions are required to prevent any exorbitance reduction. For this purpose, several constraints are provided to limit power reduction more than enough. Since the present algorithm is a multi-period optimization algorithm, there is full control on each light in all periods. Therefore, the algorithm can prevent reduction more than enough from only some particular lights in continues periods. It means the situation of each light changes during all periods by comfort constraints and priority numbers.

Figure 1 illustrates the algorithm of the present methodology with detailed steps. It should be noted that to achieve the algorithm purposes, the cooperation of all components of the agents are required for providing essential input data. Initial data such as rated power consumption of the lights, nominal power consumption of the lights, and power consumption of other existing devices in the building are the pre-optimization requirements that should be provided by SCADA system and the other agents.

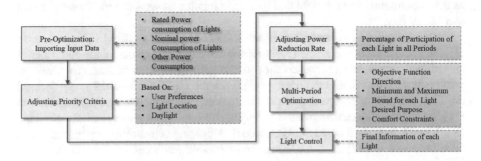

Fig. 1. The procedure of optimization algorithm.

Priority numbers are determinative parameters in optimization algorithm which can set the role of each light in the optimization. As can be seen in Fig. 1, these priority numbers are subject to several issues such as user preferences, light location, and natural light. Another parameter that makes the reduction in lights more rigid is the power reduction rate for each light in all periods. This parameter limits the power reduction in each particular light in all periods and causes the reduction to be divided into all of the lights. After defining these parameters, variables should be bounded and the relative constraints should be defined. The desired purpose of this optimization algorithm is reducing specified power reduction with observing all the existing constraints. This desired power reduction can be determined according to several aspects such as electricity price variation, ON-Peak or OFF-Peak hours, rated power generation, and energy storages if exist. After specifying all the required data, the algorithm runs, and the results are visible.

The proposed methodology is defined as a Linear Programming (LP) optimization problem, which is modeled via "OMPR" package of Rstudio® (www.rstudio.com) and is solved via "GLPK" library.

The Objective Function (OF) of the proposed optimization algorithm is shown in (1) in order to minimize the power consumption of the lights.

$$Minimize \ OF = \sum_{t=1}^{T} \sum_{l=1}^{L} Priority_{(l,t)} \times P_{(l,t)} \tag{1}$$

Priority is the number between 0 and 1 that is dedicated to each light for representing the importance of each light for the users and the bigger priority numbers are allocated to more important lights. P is the decision variable of the algorithm that shows the amount of power that should be reduced from each light in each period. It should be noted that T and L are the maximum number of periods and lights respectively.

The definition of upper bounds related to the amount of power reduction and priority of each light in each period are developed in the scope of the author's previous work [12], and they are not mentioned in this section.

Equation (2) is modeled to show the total power reduction in all the lights in each period. The objective function in (1) is subject to (2) and (3).

$$\sum_{l=1}^{L} P_{(l,t)} = RR_{(t)} \tag{2}$$
$$\forall t \in \{1, \ldots, T\}$$

RR is an abbreviation of Required Reduction in each period from all the lights. Equation (3) makes each light restricted individually by the power reduction rate coefficient in order to maintain user comfort. It means the total power reduction of each light in all periods can be adjusted and limited by (3).

$$\sum_{t=1}^{T} P_{(l,t)} = PRR_{(l)} \times \sum_{t=1}^{T} init.P_{(l,t)} \tag{3}$$
$$\forall l \in \{1, \ldots, L\}$$

PRR is brevity of Power Reduction Rate for light. *PRR* can be as a percentage of total actual power consumption of each light in all periods. The initial power consumption of each light is shown by *init.P* that means the power consumption of light in a normal situation and without any reduction. Also, as it was shown on (3), the power reduction of each light in all periods cannot exceed a defined limitation for observing user comfort. PRR can make power reduction rigid in several ways. It can be used as a coefficient for each light individually. It also can be also as a function of time. For instance, it can limit the power reduction of certain lights in certain periods of time.

3   Case Study and Results

An office building is considered that its illumination system based on fluorescent lights with DALI ballasts. The building includes 8 offices and one corridor. Each office has two 100 W lights, and the corridor contains four 100 W lights. The present study considers 20 controllable and reducible lights. Figure 2 illustrates the plan of the building.

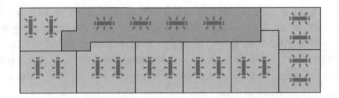

Fig. 2. Plan of the office building with 20 lights.

The total consumption of the lighting system in the building will be 2000 W. However, the minimum reduction for each light is supposed to equal to zero, while the maximum reduction stays for 65% of nominal consumption of the light. It should be noted that, in order to avoid turning off any light completely, the maximum reduction is bounded. The algorithm surveys from 8 am to 8 pm with 15 min time intervals. According to the day-ahead data, the initial consumption of the lights and non-controllable consumption of the building is shown in Fig. 3 for 48 periods. The Required Reduction (RR), can be determined based on diverse issues such as difference in production and generation rate in the building, or existing power in energy storages.

According to Fig. 3, the power consumption of the lights has been varied during the day and in periods 47 and 48 all the lights have been turned off. Since the Power Reduction Rate (PRR) is an essential parameter that impresses user comfort directly, in each execution, different values of PRR are considered in order to validate the impact of comfort constraints more precisely. The result of the first execution are in Fig. 4 while the constraint demonstrated by (3) is ignored. As seen in Fig. 4(A), the power reduction in some lights, such as L7, L9, and L12, is much more than the other ones. Therefore, Fig. 5 shows the output of the next implementation of algorithm while PRR defined to 35% for all the lights.

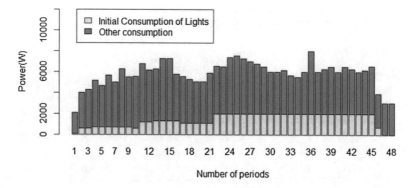

Fig. 3. The classified initial consumption of the building.

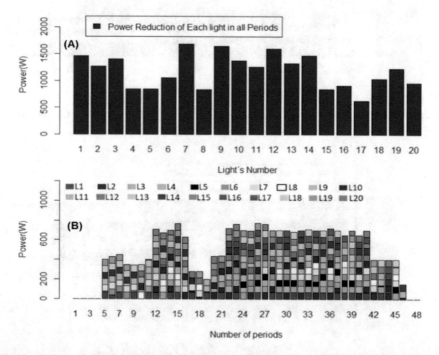

Fig. 4. Optimization results without considering user comfort constraint; (A) sum of power reduction of each light in all periods, (B) power reduction from all lights in each period.

According to Fig. 6, power reduction in each light has been slightly changed when PRR is equal to 50%. By comparing the results in this section, the importance of the parameter PRR is obvious. In this case, while the PRR is on 35%, the optimization algorithm has the best performance, somehow, it reduced the amount of power reduction by maintaining a high level of user comfort. The balanced cooperation of the lights is obviously visible in Fig. 5 while the required reduction is constant. Figure 6 presents obtained results when PRR is equal to 50%.

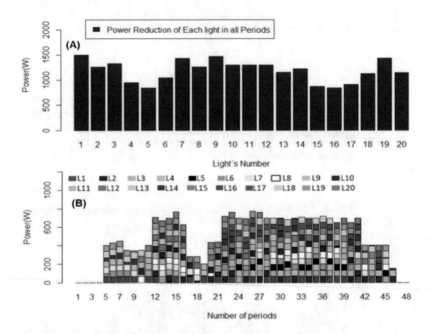

Fig. 5. Optimization results with 35% user comfort level; (A) sum of power reduction of each light in all periods, (B) power reduction from all lights in each period.

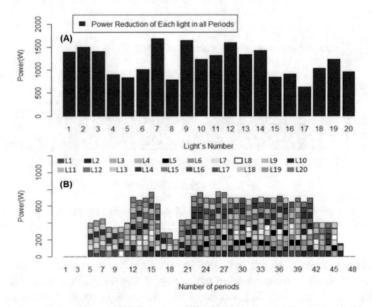

Fig. 6. Optimization results with 50% user comfort level; (A) sum of power reduction of each light in all periods, (B) power reduction from all lights in each period.

4   Conclusions

Due to the high penetration of the buildings in energy consumption, the use of optimization algorithms plays a key role. Therefore, all the producers and prosumers should be equipped with the automation infrastructures as well as intelligent decision algorithms, in order to perform the management programs, like demand response.

A multi-period optimization algorithm has been proposed in this paper to minimize the energy consumption of the lights with considering user comforts. In addition to the definition of lights priorities, a power reduction rate parameter has been defined with a direct impact on user comfort. The algorithm was implemented in a multi-agent Supervisory Control and Data Acquisition system of an office building. The focus of this paper was given to the Optimizer agent, where the developed algorithm was executed in order to optimize the consumption of the lighting system in the building.

In the case study of this work, the impact of the developed optimization algorithm is demonstrated during the working hours. A key parameter in the user comfort constraint was changed in three levels, in order to validate the performance and select the most appropriate value for this case. The obtained results of the algorithm demonstrated and proved that how the used constraints made a balance on power reduction and user comfort level.

References

1. Chen, Y., Xu, P., Gu, J., Schmidt, F., Li, W.: Measures to improve energy demand flexibility in buildings for demand response (DR): a review. Energy Build. **177**, 125–139 (2018)
2. Abrishambaf, O., Faria, P., Vale, Z.: SCADA office building implementation in the context of an aggregator. In: IEEE 16th International Conference on Industrial Informatics (INDIN), Porto, Portugal (2018)
3. Faria, P., Vale, Z.: Demand response in electrical energy supply: an optimal real time pricing approach. Energy **36**(8), 5374–5384 (2011)
4. Cao, Y., Du, J., Soleymanzadeh, E.: Model predictive control of commercial buildings in demand response programs in the presence of thermal storage. J. Clean. Prod. **218**, 315–327 (2019)
5. Althaher, S., Mancarella, P., Mutale, J.: Automated demand response from home energy management system under dynamic pricing and power and comfort constraints. IEEE Trans. Smart Grid **6**(4), 1874–1883 (2015)
6. Nguyen, D., Le, L.: Joint optimization of electric vehicle and home energy scheduling considering user comfort preference. IEEE Trans. Smart Grid **5**(1), 188–199 (2014)
7. Kandasamy, N., Karunagaran, G., Spanos, C., Tseng, K., Soong, B.: Smart lighting system using ANN-IMC for personalized lighting control and daylight harvesting. Build. Environ. **139**, 170–180 (2018)
8. Khorram, M., Abrishambaf, O., Faria, P., Vale, Z.: Office building participation in demand response programs supported by intelligent lighting management. Energy Inform. **1**(1), 9 (2018)

9. Santos, G., Femandes, F., Pinto, T., Silva, M., Abrishambaf, O., Morais, H., Vale, Z.: House management system with real and virtual resources: energy efficiency in residential microgrid. In: Global Information Infrastructure and Networking Symposium (GIIS), Porto, Portugal, pp. 1–6 (2016)
10. Gazafroudi, A., Pinto, T., Prieto-Castrillo, F., Corchado, J., Abrishambaf, O., Jozi, A., Vale, Z.: Energy flexibility assessment of a multi agent-based smart home energy system. In: IEEE 17th International Conference on Ubiquitous Wireless Broadband (ICUWB), Salamanca, Spain, pp. 1–7 (2017)
11. Khorram, M., Faria, P., Abrishambaf, O., Vale, Z.: Demand response implementation in an optimization based SCADA model under real-time pricing schemes. In: Advances in Intelligent Systems and Computing, pp. 21–29 (2019)
12. Khorram, M., Faria, P., Abrishambaf, O., Vale, Z.: Lighting consumption optimization in an office building for demand response participation. In: Power Systems Conference Clemson University (PSC), Charleston, SC, USA (2018)
13. Ogunjuyigbe, A., Ayodele, T., Akinola, O.: User satisfaction-induced demand side load management in residential buildings with user budget constraint. Appl. Energy **187**, 352–366 (2017)

Special Session on AI–Driven Methods for Multimodal Networks and Processes Modeling (AIMPM)

Special Session on AI–Driven Methods for Multimodal Networks and Processes Modeling (AIMPM)

The special session entitled **AI–driven methods for Multimodal Networks and Processes Modeling** (AIMPM 2019) is a forum that will share ideas, projects, researches results, models, experiences, applications etc. associated with artificial intelligence solutions for different multimodal networks born problems (arising in transportation, telecommunication, manufacturing and other kinds of logistic systems). The session will be held in Ávila as the part of **The 16th International Symposium Distributed Computing and Artificial Intelligence 2019**.

Recently a number of researchers involved in research on analysis and synthesis of Multimodal Networks devote their efforts to modeling different, real-life systems. The generic approaches based on the AI methods, highly developed in recent years, allow to integrate and synchronize different modes from different areas concerning: the transportation processes synchronization with concurrent manufacturing and cash ones or traffic flow congestion management in wireless mesh and ad hoc networks as well as an integration of different transportations networks (buses, rails, subway) with logistic processes of different character and nature (e.g., describing the overcrowded streams of people attending the mass sport and/or music performance events in the context of available holiday or daily traffic services routine). Due to the above mentioned reasons the aim of the workshop is to provide a platform for discussion about the new solutions (regarding models, methods, knowledge representations, etc.) that might be applied in that domain.

Organization

Organizing Committee

Chairs

Peter Nielsen   Aalborg University, Denmark
Paweł Sitek   Kielce University of Technology, Poland
Sławomir Kłos   University of Zielona Góra, Poland
Grzegorz Bocewicz   Koszalin University of Technology, Poland

Co-chairs

Izabela E. Nielsen   Aalborg University, Denmark
Zbigniew Banaszak   Koszalin University of Technology, Poland
Paweł Pawlewski   Poznan University of Technology, Poland
Mukund Nilakantan   Aalborg University, Denmark
Robert Wójcik   Wrocław University of Technology, Poland
Marcin Relich   University of Zielona Gora, Poland
Arkadiusz Gola   Lublin University of Technology, Poland
Justyna Patalas-Maliszewska University of Zielona Góra, Poland

Capacitated Vehicle Routing Problem with Pick-up, Alternative Delivery and Time Windows (CVRPPADTW): A Hybrid Approach

Paweł Sitek, Jarosław Wikarek[✉],
and Katarzyna Rutczyńska-Wdowiak

Department of Control and Management Systems,
Kielce University of Technology, Kielce, Poland
{sitek,j.wikarek,k.rutczynska}@tu.kielce.pl

Abstract. The Capacitated Vehicle Routing Problem with Pick-up, Alternative Delivery and Time Windows (CVRPPADTW) is discussed in the paper. The development of this problem was motivated by postal items distribution issues. In some approximation, one can say that the problem under consideration is a combination of many variants of the classical VRP, such as CVRP (Capacitated Vehicle Routing Problem), VRPPD (Vehicle Routing Problem with Pickup and Delivery) and VRPTW (Vehicle Routing Problem with Time Windows). What makes it different is the introduction of alternative delivery points and parcel lockers incorporated into the distribution network. The original hybrid approach integrating CLP (Constraint Logic Programming) and MP (Mathematical Programming) was used for the modeling and solving of the problem.

Keywords: Capacitated vehicle routing problem ·
Constraint logic programming · Mathematical programming · Hybrid methods ·
Optimization

1 Introduction

Rapid development of information technologies, in particular mobile technologies and on-line Internet services has substantially increased the importance of the e-commerce segment. According to the Eurostat surveys, the percentage of people aged between 16 and 74 in the EU-28 who have been ordering or buying goods or services online for private needs is still increasing. In 2016, it reached 55%, an increase of 11% compared to 2012. In 2016, about three-quarters of the residents of the Netherlands, Germany and Sweden ordered or bought goods or services via the Internet, and this percentage was even higher in Luxembourg (78%), Denmark (82%) and the UK (83%). Lower percentages were recorded in Italy and Cyprus (below 30%), with the lowest in Bulgaria (17%) and Romania (12%) [1]. The continuous growth of online shopping results in the development and growing importance of the e-commerce market within the global economy. This leads to the increasing demand for postal and courier services and raises the number of distribution companies that specialize in supplying online ordered goods.

© Springer Nature Switzerland AG 2020
E. Herrera-Viedma et al. (Eds.): DCAI 2019, AISC 1004, pp. 33–40, 2020.
https://doi.org/10.1007/978-3-030-23946-6_4

Suppliers develop and implement various innovative solutions. These include shipment tracking mechanisms, B2B and B2C platforms and a new type of delivery points in the distribution network - parcel lockers. Both including the parcel lockers in the distribution network and, due to their limited capacity, the introduction of alternative delivery points were studied and reported in [2], where the original model for a capacitated vehicle routing problem with pick-up and alternative delivery (CVRPPAD) was proposed. The aim of this study is to extend the functionality of the model proposed in [2] to include the so-called time windows [3].

2   Capacitated Vehicle Routing Problem with Pick-up, Alternative Delivery and Time Windows (CVRPPADTW): Problem Statement

The CVRPPADTW can be described as a variant of VRP [4–6] with additional possibilities of having means of transportation (vehicles) for delivery and pick-up. The capacities of the vehicles and some of the customers are taken into account. Each customer represents a post office, an individual customer or a parcel locker of a specified capacity at a given time. The problem assumes alternative delivery points, a characteristic feature of parcel locker locations. Vehicles/couriers are universal in that they can serve any customer type and do pickups and deliveries. The delivery process includes the so-called time windows, i.e., time intervals within which each customer has to be served. Considering the above, the problem can be defined as the CVRPPADTW (Capacitated Vehicle Routing Problem with Pick-up, Alternative Delivery and Time Windows), an extension of the CVRPPAD proposed in [2].

The objective is to minimize the distances the couriers have to travel and the "penalty" for delivering items to alternative points. This is the most common objective function chosen for practical reasons, but other options are possible, for example, the shortest delivery time or the number of trips/couriers, etc.

Problem characteristics (main features):

- A given set of delivery points (customers) is divided into three subsets, i.e., post offices, individual customers, parcel lockers.
- The fleet of vehicles is heterogeneous.
- A vehicle may perform no more than one trip in a planning period.
- Vehicles' capacity constraints are introduced.
- Parcel lockers' capacity constraints are introduced.
- There is a possibility of alternative delivery points.
- Each route between the points has a specified travel time and the overall travel time length cannot be greater than the set travel time length.
- There is no possibility of reloading the items.
- The parcels/letters can be both delivered to and collected from a given point.
- Each courier's route starts from and ends at the hub.
- Time windows' are introduced.

The *route length, number of points on the route, prioritize customers or shipments* constraints are not introduced in this problem, however the model is flexible enough to add these constraints if needed.

The distribution network for the CVRPPADTW is shown in Fig. 1. The main constraints of the CVRPPADTW are the same as for the CVRPPAD, as described in Table 1. The model has to be extended to include the time windows. Parameters *Time_min, Time_max*, decision variables $U_{c,p,i}$, $F_{c,p,i}$ and constraints (ctw1)…(ctw7) are added to the CVRPPAD model [2], as summarized in Table 2.

$$\text{Time\_min}_i \leq \sum_{c \in C} \sum_{j \in P \cup d} F_{cpi} \leq \text{Time\_max}_i \ \forall i \in I, \qquad \text{(cwt1)}$$

$$U_{cpi} \leq F_{cpi} \leq A \cdot U_{cpi} \ \forall i \in I, c \in C, p \in P \qquad \text{(cwt2)}$$

$$U_{cpi} \leq \sum_{j \in P} Y_{cpji} \leq A \cdot U_{cpi} \forall i \in I, c \in C, p \in P \qquad \text{(cwt3)}$$

$$A \cdot U_{cpi} \geq \sum_{j \in P} Y_{cpji} \forall i \in I, c \in C, p \in P \qquad \text{(cwt4)}$$

$$F_{cpi} + ti_{p,j} = F_{cpj} \ \forall i \in I, c \in C, p \in P, j \in P \forall Y_{cpji} = 1 \qquad \text{(cwt5)}$$

$$U_{cpi} = \{0, 1\} \ \forall c \in C, p \in P \cup d, i \in I \qquad \text{(cwt6)}$$

$$F_{cpi} \in Z^{+} \ \forall c \in C, p \in P \cup d, i \in I \qquad \text{(cwt7)}$$

Table 1. The main constraints (C) of the CVRPPAD

Name	Description
c1	Arrival and departure of a courier/vehicle at/from the delivery point (parcel locker, post office, and customer)
c2	If no items are to be carried on the route, a courier/vehicle does not travel that route
c3	If a courier/vehicle does not travel along a route, no items are to be carried on that route
c4	At no route segment courier carries more items than allowable vehicle volume
c5	Item being delivered to one of possible delivery points
c6	One trip of a courier/vehicle
c7	Item delivered to only one delivery point
c8	Not to exceed a point volume/At the point – the number of items matches the number of vacancies/
c9	Items picked up/delivered from/to a delivery point
c10	Trips completed within the required time
c11	Each courier picked up/delivered an item from/to a source (hub)
c12	Which courier carries a given item
c13	Only one courier carries a given item
c14	Binarity

Table 2. Parameters, decision variables and constraints

Name	Description
Parameters	
Time_min,	The earliest time moment of courier's arrival
Time_max	The latest time moment of courier's arrival
Decision variables	
$U_{c,p,i}$	If courier c delivers shipment p to point i then $U_{c,p,i} = 1$, otherwise $U_{c,p,i} = 0$
$F_{c,p,i}$	The time moment at which courier c delivers consignment p to point i ($F_{c,p,i} = 0$ if courier c does not deliver shipment i to point p)
Constraints	
cwt1	The delivery time must be within the specified time interval
cwt2	If the shipment i is not delivered to point p, then the delivery time is zero (linked decision variables $U_{c,p,i}$ and $F_{c,p,i}$)
cwt3	If the shipment i is delivered to point p, then non-zero delivery time must be specified (linked decision variables)
cwt4	Defines delivery time at subsequent points
cwt5	The delivery time must be within the specified time interval
cwt6, cwt7	Binarity

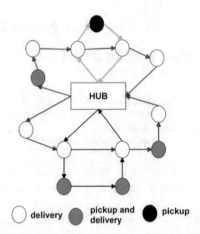

Fig. 1. Example distribution network for CVRPPADTW.

3   Implementation

Due to the binary linear programming (BLP) character of the proposed model, the mathematical programming (MP) environment is a natural choice. A number of MP solvers are available: LINGO [7], CPLEX, SCIP, Gurobi [8], etc. but their effectiveness is low due to the classification of all VPRs as NP-hard. One of the ways to handle this computational complexity is the use of dedicated heuristic methods or metaheuristics [9], which are approximate methods. We propose a hybrid approach to the modeling and solving CVRPPADTWs, which is an original framework that integrates CLP and MP environments and enables the transformation of the problem. The transformation serves as a presolving method used to reduce the number of decision variables and constraints in the problem, thus reducing the space of potential solutions. The hybrid approach has been the subject of numerous studies [2, 10–14]. The general idea of the hybrid approach for the CVRPPADTW is shown in Fig. 2.

For the implementation purposes, Eclipse (implementing part of the CLP) and interchangeably LINGO and Gurobi (implementing part of the MP) are used.

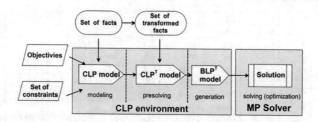

Fig. 2. The concept of implementation for hybrid approach.

4 Computational Experiments

Many computational experiments were carried out to verify the correctness of the CVRPPADTW model and to test the effectiveness of the hybrid approach. In the first phase of experiments, calculations were done for examples P1…P10, which differed in the number of shipments (Lpr), the number of delivery points (Lpu) and the number of couriers/vehicles (Lku). In these experiments, time windows were not taken into consideration, so the model from [2] was solved. A novelty was the use of an MP–Gurobi [8] efficient solver instead of LINGO. In the second phase of experiments, for Example P5, the time windows (TW) were introduced for selected delivery points (Table 3). In both phases, the experiments were carried out for the same data sets and

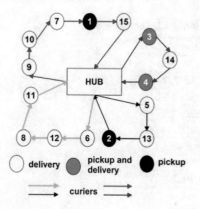

Fig. 3. Distribution network for example P5 without TW.

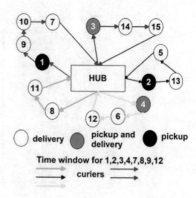

Fig. 4. Distribution network for example P5 with TW.

the same examples using the MP-based and hybrid approaches. The results are presented in Table 4. Additionally, the results for example P5 without and with time windows are shown in the form of distribution network graphs in Figs. 3 and 4 respectively.

Table 3. Results of computational experiments

P	Lpr	Lpu	Lku	MP-based approach				Hybrid approach(CLP and MP)			
				V_{BIN}	C	T	F_C	V_{INT}	C	T	F_C
MP solver GUROBI											
P1	20	8	3	1834	478	123	156	596	302	1	156
P2	20	8	3	1834	478	234	42	596	302	3	42
P3	20	8	3	1834	478	345	56	596	302	8	56
P4	20	15	6	5453	934	123	245	2946	2124	41	245
P5	40	15	6	11567	1343	234	323	2946	2124	45	323
P6	60	15	6	17456	1678	345	323	2946	2124	56	323
P7	80	15	6	21234	1945	456	435	2946	2124	89	435
P8	10^2	15	6	26789	2134	567	467	2946	2124	234	467
P9	10^3	100	20	35×10^6	21034	2400*	NF	273456	467545	823	58687
P10	10^3	200	20	56×10^6	45567	2400*	NF	466434	794534	1234	103423
P	Lpr	Lpu	Lku	MP-based approach				Hybrid approach(CLP and MP)			
				V_{BIN}	C	T	F_C	V_{INT}	C	T	F_C
MP solver GUROBI											
P5_1	40	15(1)	6	18767	2063	254	386	3167	2568	47	386
P5_2	40	15(3)	6	18767	2063	267	423	3456	2845	56	423
P5_3	40	15(4)	6	18767	2063	278	456	3678	3134	67	456
P5_4	40	15(8)	6	18767	2063	345	834	3945	3342	78	834

Lpu The number of points (with time window)
F_c The optimal value of objective function
V_{BIN} The number of binary decision variables
V_{INT} The number of integer decision variables
* Calculations interrupted after time 2400 s
NF No feasible solution found
Lpr The number of items
T Computational time in seconds
Pi Example
Lku The number of couriers

Table 4. The time windows for example 5

Customer	Time
1	7:30–8:30
2	7:30–8:30
3	7:30–8:30
4	7:30–8:30
7	10:00–11:00
8	7:30–8:30
9	8:00–9:00
12	9:30–10:30

5 Conclusion

Analysis of the results leads to the following conclusions:

- The introduction of time windows increases the computing time both in the MP-based approach and in the hybrid approach.
- The increase in computing time depends on the number of time windows.
- The incorporation of time windows makes a larger number of couriers be involved in the distribution process for the same consignment volume.
- Application of the hybrid approach for solving both problems (without and with time windows) substantially reduces the computing time of more than a hundredfold.
- No solution in the acceptable time was found for the larger-size examples (P9, P10) in the MP-based approach.
- MP solver Gurobi is far more effective than LINGO.

Further studies will focus on the introduction of other objective functions for the CVRPPPADTW and constraints on *route length, number of points on the route, prioritize customers or shipments, lead times* [15], etc. Logical constraints related to business, legal conditions, etc. and fuzzy logic [16, 17] will also be added to the model.

Next the results of CVRPPADTW (capacitated vehicle routing problem with pick-up, alternative delivery and time windows) with the use of hybrid approach will be compared with the results of search for the solution with the use of genetic algorithm (GA). The use of GA is justified due to the large number of constraints and decision variables. The genetic algorithms have a stochastic character, and so they do not guarantee obtaining optimum solution, but it is expected, the best individual will represent solution nearing the optimum one. In work [18] it was shown the effectiveness and efficiency of the genetic algorithms depend on genetic representation of individuals, genetic operations, stop criterion and the set of control parameters. The algorithm should be appropriately designed to prevent premature convergence and to guarantee the shortest possible computation time. It is also planned to hybridize the declarative approach with other meta-heuristics, such as shuffled frog-leaping algorithm (SFLA) [19], Ant colony optimization (ACO) etc.

References

1. Eurostat - Statistics Explained - Europa.eu. http://ec.europa.eu/eurostat/statisticsexplained/index.php/Main_Page. Accessed May 10 2018
2. Sitek, P., Wikarek, J.: Capacitated vehicle routing problem with pick-up and alternative delivery (CVRPPAD): model and implementation using hybrid approach. Ann. Oper. Res. 1–21 (2017). doi:https://doi.org/10.1007/s10479-017-2722-x
3. Azi, N., Gendreau, M., Potvin, J.Y.: An Exact algorithm for a vehicle routing problem with time windows and multiple use of vehicles. Eur. J. Oper. Res. **202**(3), 756–763 (2010). https://doi.org/10.1016/j.ejor.2009.06.034

4. Jairo, R., Montoya, T., Francob, J.L., Isazac, S.N., Jiménezd, H.F., Herazo-Padillae, N.: A literature review on the vehicle routing problem with multiple depots. Comput. Ind. Eng. **79**, 115–129 (2015)

5. Kumar, S.N., Panneerselvam, R.: A survey on the vehicle routing problem and its variants. Intell. Inf. Manag. **4**, 66–74 (2012)

6. Wassan, N., Wassan, N., Nagy, G., Salhi, S.: The multiple trip vehicle routing problem with backhauls: formulation and a two-level variable neighbourhood search. Comput. Oper. Res. (2016). https://doi.org/10.1016/j.cor.2015.12.017

7. Lindo. http://www.lindo.com/. Accessed May 04 2018

8. Gurobi. http://www.gurobi.com/. Accessed May 04 2018

9. Archetti, C., Speranza, M.G.: A survey on matheuristics for routing problems. EURO J. Comput. Optim. **2**(223) (2014). https://doi.org/10.1007/s13675-014-0030-7

10. Wikarek, J.: Implementation aspects of hybrid solution framework. In: Recent Advances in Automation, Robotics and Measuring Techniques, vol. 267, pp. 317–328 (2014). https://doi.org/10.1007/978-3-319-05353-0_31

11. Sitek, P., Wikarek, J.: A multi-level approach to ubiquitous modeling and solving constraints in combinatorial optimization problems in production and distribution. Appl. Intell. **48**, 1344–1367 (2018). https://doi.org/10.1007/s10489-017-1107-9

12. Bockmayr, A., Kasper, T.: A framework for combining CP and IP, branch-and-infer, constraint and integer programming. Toward Unified Methodol. Oper. Res./Comput. Sci. Interfaces **27**, 59–87 (2014)

13. Milano, M., Wallace, M.: Integrating operations research in constraint programming. Ann. Oper. Res. **175**(1), 37–76 (2010)

14. Sitek, P., Wikarek, J., Nielsen, P.: A constraint-driven approach to food supply chain management. Ind. Manag. Data Syst. **117**, 2115–2138. https://doi.org/10.1108/imds-10-2016-0465

15. Nielsen, P., Michna, Z., Do, N.A.D.: An empirical investigation of lead time distributions. In: IFIP Advances in Information and Communication Technology, vol. 438 (PART 1), pp. 435–442 (2014). https://doi.org/10.1007/978-3-662-44739-0_53

16. Kłosowski, G., Gola, A., Świć, A.: Application of fuzzy logic in assigning workers to production tasks. In: Omatu, S., Selamat, A., Bocewicz, G., Sitek, P., Nielsen, I., Garcia-Garcia, J.A., Bajo, J. (eds.) Distributed Computing and Artificial Intelligence, 13th International Conference, Advances in Intelligent Systems and Computing, vol. 474, pp. 505–513. Springer (2016)

17. Bocewicz, G., Banaszak, Z., Nielsen, I.: Multimodal processes prototyping subject to grid-like network and fuzzy operation time constraints. Ann. Oper. Res. (2017). https://doi.org/10.1007/s10479-017-2468-5

18. Rutczyńska-Wdowiak, K.: Replacement strategies of genetic algorithm in parametric identification of induction motor. In: 22nd International Conference On Methods And Models In Automation And Robotics (mmar), pp. 971–975 (2017). https://doi.org/10.1109/mmar.2017.8046961

19. Crawford, B., Soto, R., Pena, C., Palma, W., Johnson, F., Paredes, F.: Solving the set covering problem with a shuffled frog leaping algorithm. In: Nguyen, NT., Trawinski, B., Kosala, R. (eds.) Intelligent Information and Database Systems. Lecture Notes in Artificial Intelligence, vol. 9012, pp. 41–50 (2015). https://doi.org/10.1007/978-3-319-15705-4_5_5

Predicting the Error of a Robot's Positioning Repeatability with Artificial Neural Networks

Rafał Kluz[1] , Katarzyna Antosz[1] , Tomasz Trzepieciński[1],
and Arkadiusz Gola[2(✉)]

[1] Rzeszow University of Technology, Rzeszow, Poland
{rkktmiop, katarzyna.antosz, tomtrz}@prz.edu.pl
[2] Lublin University of Technology, Lublin, Poland
a.gola@pollub.pl

Abstract. Industrial robots are an integral part of modern manufacturing systems. In order to fully use their potential, the information related to the robot's accuracy should be known first of all. In most cases, the information considering robot's errors, provided in a technical specification, is scarce. That's why, this paper presents the issues of determining the error of industrial robots positioning repeatability. A neural mathematical model that allows for predicting its value with the error less than 5% was designed. The obtained results were compared to a classical mathematical model. It was revealed that a well-trained neural network enables the prediction of the error of positioning repeatability with the doubled accuracy.

Keywords: Robot · Assembly stand · Positioning repeatability ·
Neural networks

1 Introduction

Modern concepts considering production are characterized by the integrity of machines, systems and the introduction of changes in manufacturing processes aiming at the increase of production efficiency with the possibility of flexible changes of a production range [1–3]. Thus, these concepts are inherently related to the production process robotisation. Industrial robots have been successfully used in a number of industry branches, e.g. for painting, packaging or manipulating. However, using them for more accurate processes still encounters many problems [4, 5]. These processes require high accuracy of robots (positioning repeatability and accuracy).

Frequent changes of a production range demand from a user the knowledge of a robot's error value. Nevertheless, assuming a maximum error value, a user often underuses his/her stand. Then, it is necessary to measure the error value in a certain point in its space. For this purpose, a number of methods [6, 7] may be used. They are characterized by different labour consumption of measurements and the accuracy of the results obtained. In practice, the models based on a robot's kinematics are most often used. They allow to determine the error of positioning repeatability on the basis of the knowledge of extreme values of $\pm \Delta q_i$ configuration coordinates setup error, or on the basis of variational summation of random variables caused by the errors of a setup of

© Springer Nature Switzerland AG 2020
E. Herrera-Viedma et al. (Eds.): DCAI 2019, AISC 1004, pp. 41–48, 2020.
https://doi.org/10.1007/978-3-030-23946-6_5

these coordinates [6, 8]. However, the use of these methods requires thorough research and an accurate estimation of the errors of a configuration coordinates setup. On account of the issue complexity, the models are in many cases burdened with a considerable error, what is reflected in the accuracy of the realized processes. There may be an alternative solution to the presented problem, that is using artificial neural networks which do not need to model constituent errors because they are mainly based on the results of experimental research. That's why, this work attempts to design a numerical model that uses an artificial neural network in order to predict the value of a robot's positioning repeatability error.

The second chapter of the work presents the mathematical model of repeatability error of the industrial robot and the results of experimental researches. The third presents the developed of neural networks model. The last chapter compares the results of experimental studies and the classical mathematical model with the results of the developed model.

2 Mathematical Model of a Robot's Positioning Repeatability Error

A robot's positioning repeatability error depends on many factors. It may be described in the form of the function of the density of 3D random variable probability [9].

$$
f(x, y, z) = \frac{1}{\sqrt{(2\pi)^3 \det[\lambda_{jk}]}} \exp\left[-0.5 \sum_{j=1}^{3} \sum_{k=1}^{3} \Lambda_{jk} x_j x_k\right]
\tag{1}
$$

The values of a covariance matrix coordinates $[\lambda_{jk}]$, that is a matrix of the second-level moments, may be determined on the basis of the following equations [10]:

$$
\lambda_{xx} = \sum_{i=1}^{n} \left(X'_{q_i}\right)^2 \sigma_{q_i}^2 = \left(X'_{q_1}\right)^2 \sigma_{q_1}^2 + \left(X'_{q_2}\right)^2 \sigma_{q_2}^2 + \cdots + \left(X'_{q_n}\right)^2 \sigma_{q_n}^2
\tag{2}
$$

$$
\lambda_{yy} = \sum_{i=1}^{n} \left(Y'_{q_i}\right)^2 \sigma_{q_i}^2 = \left(Y'_{q_1}\right)^2 \sigma_{q_1}^2 + \left(Y'_{q_2}\right)^2 \sigma_{q_2}^2 + \cdots + \left(Y'_{q_n}\right)^2 \sigma_{q_n}^2
\tag{3}
$$

$$
\lambda_{zz} = \sum_{i=1}^{n} \left(Z'_{q_i}\right)^2 \sigma_{q_i}^2 = \left(Z'_{q_1}\right)^2 \sigma_{q_1}^2 + \left(Z'_{q_2}\right)^2 \sigma_{q_2}^2 + \cdots + \left(Z'_{q_n}\right)^2 \sigma_{q_n}^2
\tag{4}
$$

$$
\lambda_{xy} = \sum_{i=1}^{n} X'_{q_i} Y'_{q_i} \sigma_{q_i}^2 = X'_{q_1} Y'_{q_1} \sigma_{q_1}^2 + X'_{q_2} Y'_{q_2} \sigma_{q_2}^2 + \cdots + X'_{q_n} Y'_{q_n} \sigma_{q_n}^2
\tag{5}
$$

$$
\lambda_{yz} = \sum_{i=1}^{n} Y'_{q_i} Z'_{q_i} \sigma_{q_i}^2 = Y'_{q_1} Z'_{q_1} \sigma_{q_1}^2 + Y'_{q_2} Z'_{q_2} \sigma_{q_2}^2 + \cdots + Y'_{q_n} Z'_{q_n} \sigma_{q_n}^2
\tag{6}
$$

$$\lambda_{xz} = \sum_{i=1}^{n} X'_{q_i} Z'_{q_i} \sigma^2_{q_i} = X'_{q_1} Z'_{q_1} \sigma^2_{q_1} + X'_{q_2} Z'_{q_2} \sigma^2_{q_2} + \cdots + X'_{q_n} Z'_{q_n} \sigma^2_{q_n} \qquad (7)$$

where: σ^2_{qi}, is a variance of the error of qi configuration coordinate setup, $\lambda_{xx} = \sigma^2_x$, is a variance of the x component of an error vector, $\lambda_{yy} = \sigma^2_y$, is a variance of the y component of an error vector, $\lambda_{zz} = \sigma^2_z$, is a variance of the z component of an error vector, $\lambda_{xy} = \lambda_{yx} = \text{cov}(x, y)$, is a covariance of x and y components of an error vector $\lambda_{yz} = \lambda_{zy} = \text{cov}(y, z)$, is a covariance of y and z components of an error vector, $\lambda_{xz} = \lambda_{zx} = \text{cov}(x, z)$, is a covariance of x and z components of an error vector.

During the study, the measurement of the positioning repeatability error of Mitsubishi RV-M2 robot in 32 points of its workspace was done. Compliant to ISO 9283 standard, the study was conducted at the maximum load of a robot's effector. During the measurement, a measuring cube was moved to the chosen points of the robot's space with maximum speed. The robot moved between the points following a linear trajectory, thereby forcing the movement of all configuration coordinates qi. The measurement results were recorded with inductive displacement sensors of Tesa GT 61 (repeatability (±2 s): 0.3 µm, hysteresis error (±2 s): 0.2 µm, linear coefficient of expansion: 0.09 µm/ °C). The signals from the sensors were recorded and read with TT 300 gauge by Tesa. In order to determine the parameters of a random variable of the positioning repeatability error, it was moved 100 times towards the chosen points of the robot's space, and then, the deviation value of a standard error σx, σx as well as a correlation coefficient ρ were determined. Figure 1 shows histograms of the random variable of a robot's error in x and y axis direction as well as the concentration ellipsis of probability in a chosen point of its workspace.

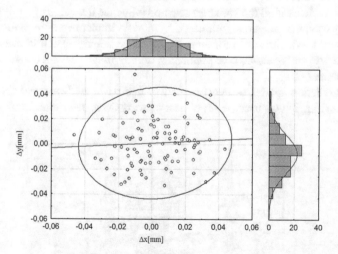

Fig. 1. The result of a robot's error measurement in a chosen point of its space

3   Neural Model of a Robot's Positioning Repeatability Error

Artificial neural networks are one of the heuristic methods, also known as "soft" calculation methods and transform information in imitation of the phenomena occurring in a human brain [11], and they may serve as a model of any object of unknown characteristics [12]. On account of the possibility of including many factors in a modelling process, neural networks allow to build a model that makes it possible to predict robot's errors with high accuracy.

The formula describing a neuron activity may be expressed as follows [13]:

$$y = f\left(\sum_{i=1}^{n} w_i u_i\right) \qquad (9)$$

where: $f(x) = 1$ if $x \geq 0$, $f(x) = 0$ if $x < 0$, $w0 = \upsilon$, $u0 = -1$, y – an output signal of a neuron, wi – a synaptic weight of i-th neuron, ui – an input signal of i-th neuron, υ - a threshold value.

In order to determine the weight and threshold values of the particular network neurons, it is necessary to prepare a learning data set which includes a set of input signal values and the corresponding output signal values. The values of configuration coordinates (qi) of an industrial robot that determine the position of an end-effector in the stand workspace and its orientation were adopted as the input signals into the network.

The expected output signal was the value of a standard deviation σx, σx of an error in x and y axis direction as well as ρ correlation coefficient. In order to calculate the neuron output value, a hyperbolic tangent function was used.

In order to check if all the variables selected influence the value of a robot's error in an essential way, the genetic algorithm (GA) module implemented in Statistica Neural Networks was used. Genetic algorithms are based on natural selection mechanisms as well as heredity; and operated on a population of individuals that are potential solutions to the problem.

As a prediction model, a network of five-neuron-input layer and three-neuron-output layer (Fig. 2) was used. A set of best networks was determined with the help of

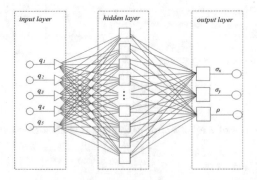

Fig. 2.   Structure of a neural network

STATISTICA Neural Networks program. Two neural networks of the best quality and a different number of neurons nu in a hidden layer: 6 neurons (ANN1) and 10 neurons (ANN2) were subjected to a comparative evaluation.

After defining the number of layers and neurons in each layer, the weight and threshold values of all neurons were selected. The learning of a perceptron network is iterative what means that in consecutive iterations (the so-called learning eras) the weights and the threshold are modified to reduce an aggregate error of a network. In the studied case, a network learning process was conducted with the help of an error backpropagation algorithm [14]. The value of an RMS error determined separately for particular data subsets was adopted as a criterion of a network quality:

$$RMS = \sqrt{\frac{\sum_{i=1}^{n} (c_i - y_i)^2}{n}} \tag{10}$$

where: n - is a number of vectors of a learning set, y_i – is a signal of an output neuron for i-th pattern, c_i – is an expected signal of an output neuron for i-th pattern.

In order to prevent a network over-learning process, the value of η indicator, which is responsible for the stability and speed of the convergence of a learning algorithm, was adopted on 0.001 level. This allows to determine the moment of completion of a network learning process (the values of an RMS error of a validation set are not further reduced).

4 The Results of Neural Networks Prediction

While evaluating a network model, a special focus should be put on the standard deviation ratio (SDR) and R Pearson correlation coefficient. SDR measurement always takes non-negative values. Its lower value indicates a better quality of the model. For a very good quality model the measurement reaches the value ranging from 0 to 0.1. If S.D. Ratio value is higher than a unit, then, using the constructed model is unjustified. That is because a more accurate evaluation of the described variable value would be its arithmetic mean determined on a learning set.

Testing the prediction of a neural learning set was conducted on the basis of four sets of the input data for which the value of a robot's error was determined experimentally. The comparison of the results of a neuronal network, experimental studies and the results obtained on the mathematical modelling are presented in Table 1.

The analysis of the results shows that ANN2 of 10 neurons in a hidden layer possesses the best predictive properties. The RMS error value of this network for a test set was respectively 0,0010 (σ_x), 0,0008 (σ_y) and 0,026 (ρ). The worst results were obtained for a mathematical model. The RMS error, in this case, was respectively 0,002 (σ_x), 0,0015 (σ_y) and 0,079 (ρ). ANN2 shows the most even values of the network prediction errors (Figs. 3, 4) for the four chosen observation vectors. The average error of the prediction for the presented network reached the value of 3.97% for the error in x axis direction, 4.66% in y axis direction and 14.31% for the correlation coefficient.

Table 1. Comparison of the prediction of a robot's repeatability using artificial neuronal networks with the results of mathematical modelling

Robot's positioning repeatability error		Number of observation sets			
		5	10	24	27
Experiment	σ_x	0,0213	0,0171	0,0141	0,0153
	σ_y	0,0194	0,0192	0,0162	0,0141
	ρ	0,154	−0,200	−0,314	0,025
ANN1	σ_x	0,0224	0,0165	0,0128	0,0163
	σ_y	0,0191	0,021	0,0172	0,0140
	ρ	0,244	−0,372	−0,581	0,009
ANN2	σ_x	0,0223	0,0166	0,0131	0,0151
	σ_y	0,0196	0,0213	0,0176	0,0144
	ρ	0,178	−0,238	−0,341	0,028
Mathematical model	σ_x	0.0201	0,0158	0,0143	0,0161
	σ_y	0,0181	0,0196	0,0183	0,0164
	ρ	0,214	−0,151	−0,452	0,045

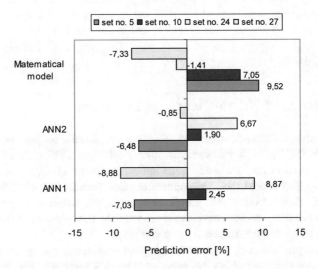

Fig. 3. Comparison of the values of the errors of a robot's positioning repeatability prediction in *x* axis direction by neural networks models.

Table 2 presents the quality measurements for the analysed networks. The analysis of the results also indicates that ANN2 possesses much better approximation properties than ANN1. The measurements for this network, calculated on the basis of a test set, show a better ability for generalization as well. During the study, the analysis of the network with 15 neurons in a hidden layer was also performed. However, high values of a standard deviation ratio (SDR) for a test set excluded the presented model. It may be explained by the fact that a set of cases used for creating and learning of the network, including 15 neurons in a hidden layer, was too small.

Table 2. The quality measurements for the analysed networks

Set		SDR			Correlation		
		σ_x	σ_y	ρ	σ_x	σ_y	ρ
Learning	ANN1	0,337	0,513	0,691	0,975	0,914	0,913
	ANN2	0,201	0,321	0,692	0,981	0,963	0,889
Validation	ANN1	0,527	0,630	0,711	0,945	0,933	0,941
	ANN2	0,430	0,491	0,513	0,966	0,971	0,863
Test	ANN1	0,436	0,552	0,751	0,996	0,934	0,941
	ANN2	0,317	0,371	0,662	0,993	0,966	0,995

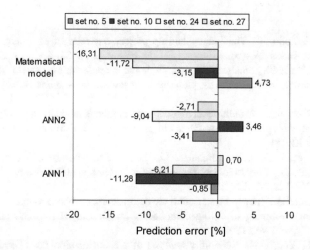

Fig. 4. Comparison of the values of the prediction of a robot's positioning repeatability error in y axis direction by neural networks models

5 Summary

Neural networks are one of the most promising modelling methods. The conducted studies connected to the prediction of a robot's positioning repeatability error provide optimistic grounds for their possible use for this type of tasks. On the basis of the numerical studies, a structure of a network that gives the possibility of the prediction of a robot's positioning repeatability error, with an average error below 3.97% in x axis direction, 4.66% in y axis direction and 14.31% for a correlation coefficient, was selected. The obtained values of the prediction of positioning repeatability in x and y axes directions are almost twice smaller than in case of a classical mathematical model (6.33% − x, 8.98% − y), and twice smaller than in case of a linear correlation coefficient. This is due to the limitations of a classical mathematical model in which it is difficult to include all the factors that influence a robot's error (yet, they are included in a network learning process).

Despite many benefits, using neural networks has its limitations. In this case, the limitation is the necessity of performing laborious experimental studies indispensable for the correct neural network training. That is why, in the real production conditions, mathematical models that require much smaller workload while determining a robot's positioning repeatability error may be equally used with the models based on neural networks despite considerably bigger prediction errors. It should be noted that the researches were carried out in laboratory conditions. The further work it should be relate to the generalization of the presented models by including in them changing conditions of the environment, which may affect both the value of the robot's error and the reliability of the process.

References

1. Terkaj, W., Tolio, T.: The italian flagship project: factories of the future. In: Tolio, T., Copani, G., Terkaj, W. (eds.) Factories of the Future, pp. 3–35. Springer, Cham (2019)
2. Gola, A.: Reliability analysis of reconfigurable manufacturing system structures using computer simulation methods. Eksploatacja i Niezawodnosc – Maint. Reliab. **21**(1), 90–102 (2019)
3. Sitek, P., Wikarek, J.: A multi-level approach to ubiquitous modeling and solving constraints in combinatorial optimization problems in production and distribution. Appl. Intell. **48**(5), 1344–1367 (2018)
4. Tsarouchi, P., Makris, S., Michalos, G., Stefos, M., Fourtakas, K., Kalsoukalas, K., Kontrovrakis, D., Chryssolouris, G.: Robotized assembly process using dual arm robot. Procedia CIRP **23**, 47–52 (2014)
5. Świć, A., Gola, A., Zubrzycki, J.: Economic optimisation of robotized manufacturing system structure for machining of casing components for electric micromachines. Actual Probl. Econ. **175**(1), 443–448 (2016)
6. Choi, D.H., Yoo, H.H.: Reliability analysis of a robot manipulator operation employing single Monte-Carlo simulation. Key Eng. Mater. **321**(323), 1568–1571 (2006)
7. Kluz, R., Kubit, A., Sęp, J., Trzepieciński, T.: Effect of temperature variation on repeatability positioning of a robot when assembling parts with cylindrical surfaces. Maint. Reliab. **20**(4), 503–513 (2018)
8. Kluz, R., Trzepieciński, T.: The repeatability positioning analysis of the industrial robot arm. Assem. Autom. **34**, 285–295 (2014)
9. Brethé, J.F., Vasselin, E., Lefebvre, D., Dakyo, B.: Modeling of repeatability phenomena using the stochastic ellipsoid approach. Robotica **24**(4), 477–490 (2006)
10. Kotulski, Z., Szczepiński, W.: Error Analysis with Applications in Engineering. Springer, Heidelberg (2009)
11. Patterson, D.W.: Artificial Neural Networks—Theory and Applications. Prentice-Hall, Englewood Cliffs (1998)
12. Trzepieciński, T., Lemu, H.G.: Application of genetic algorithms to optimize neural networks for selected tribological tests. J. Mech. Eng. Autom. **2**(2), 69–76 (2012)
13. Yegnanarayana, B.: Artificial Neural Networks. Prentice-Hall, New Delhi (2006)
14. Sivanandam, S.N., Deepa, S.N.: Introduction to Genetic Algorithms. Springer, Heidelberg (2008)

Assessing the Effectiveness of Using the MES in Manufacturing Enterprises in the Context of Industry 4.0

Małgorzata Skrzeszewska and Justyna Patalas-Maliszewska(✉)

Institute of Computer Science and Production Management,
University of Zielona Góra, Zielona Góra, Poland
m.skrzeszewska@wm.uz.zgora.pl,
J.Patalas@iizp.uz.zgora.pl

Abstract. Industry 4.0 can be defined as the next level of manufacturing enterprise development and refers e.g. to investments in digital technology. To adapting the enterprise to the Industry 4.0 concept, as the first step, the assessment of currently used information technologies (so called basis for investments in "smart" technologies) in the company should be made. The tool, that supports the execution of business processes within a company may be the information system, namely Manufacturing Execution System (MES). The use of MES enable date and information storage in real time from production business processes and their transfer to other activities realised within a company. The aim of the article is to analyse the effectiveness of using the MES by employees in maintenance departments, on the basis of two examples of manufacturing companies in the automotive sector of industry; analysis will be undertaken on three levels of management: operational, tactical and strategic. Conducting research is of particular importance, as it indicates the degree of a company's readiness to implement the Industry 4.0 Concept.

Keywords: Manufacturing Execution System (MES) ·
Manufacturing enterprise · Industry 4.0

1 Introduction

Introducing changes in computerization and automation into an enterprise, in the context of the Industry 4.0 Concept, demands, on the one hand, a detailed recognition of a company's information requirements; on the other hand, it demands that responsible work, in forecasting the efficiency of the outlays, incurred in an investment, be carried out.

The Industry 4.0 Concept has been known of since 2011 [1]. Its basic element is the need to develop information technology, automation and the robotisation of processes taking place within a company. However, investing in the digital transformation of an enterprise is expensive, hence the need to develop models and methods, the use of which will allow the effectiveness of the IT solutions implemented to be estimated.

The present article attempts to evaluate the effectiveness of the implementation of the Manufacturing Execution Systems (MES), as exemplified by the maintenance

© Springer Nature Switzerland AG 2020
E. Herrera-Viedma et al. (Eds.): DCAI 2019, AISC 1004, pp. 49–56, 2020.
https://doi.org/10.1007/978-3-030-23946-6_6

department of a production enterprise. The analysis was carried out in two, automotive industry manufacturing companies, with the research covering employees at the tactical and operational strategic levels, conducting their activities with the support of the MES.

2 The Development of Enterprises in the Context of Industry 4.0

Implementing solutions in the context of Industry 4.0, such as Intelligent Factories, the Internet of Things or autonomous robots, has, as its aim, a reduction in costs, including maintenance costs in plants, resulting in flexible, dispersed and ready-made production, tailored to the needs of customers in real time. One feature of Industry 4.0 is robotisation. The new, 'World Robotics' report shows that in 2017, a record number of 381,000 units were delivered worldwide, an increase of 30%, compared to the previous year. This means that the annual sales volume of industrial robots has increased by 114% over the last five years (2013–2017) and that the value of sales has increased by 21%, compared to 2016 (Fig. 1).

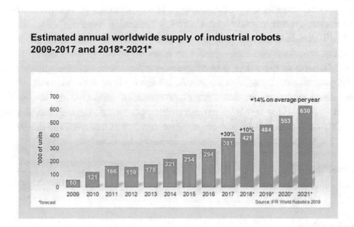

Fig. 1. Estimated global supply of industrial robots in 2009–2017 with the forecasts for 2018–2019.

A company's activity, according to Industry 4.0, allows more flexible work organization models to be initiated in order to meet the growing needs of employees. It makes possible the balance between work and private life, as well as between personal development and on-going professional development [2]. In today's extremely competitive environment, manufacturing companies must withstand growing, global competition in a variety of strategic dimensions, such as product innovation, quality, production costs, the speed of making good decisions and the flexibility of production processes. [3] indicated that according to Industry 4.0 the flexible manufacturing systems (FMS) for the manufacturing companies should be developed. In the Industry 4.0 Concept, the most important element of any production is the ability to access data and information in real time; this is now possible with the use of the Manufacturing Execution Systems (MES) for production management.

The MES system supports management, the monitoring and synchronisation of real-time physical processes, related to the processing of raw materials into semi-finished products and/or finished products. They co-ordinate the implementation of work orders with the production schedule and systems at the enterprise level. MES applications also provide feedback on process performance and, where necessary, support traceability, genealogy and integration with the history of processes at the component and material levels. The MES system also performs functions characteristic of the SCADA system (Supervisory Control and Data Acquisition), a unified interface for production controllers and autonomous industrial systems.

Assessing efficiency levels requires effectiveness of measurement. Measurement of effectiveness is one of the most dynamically developing concepts of recent years. Many methods and techniques have been created to ensure effective measurement and adequate results, which will then be translated into appropriate, corrective actions and improve the functioning of the organisation [4]. In the literature on the subject, according to [5], the assessment of the effectiveness of ICT's should be multi-dimensional and cannot be separated, either from the context, or from the circumstances of the organisation. However, according to [6] the principle of efficiency is characteristic for all activities through which some rational goals are attempted. We state, that one synthetic criterion of an organisation's efficiency cannot be introduced, nor can it be possible to formulate a single, universally, multi-rated evaluation system. Such a system must be constructed individually, taking into account the genetic type of an organisation's function, the set of current goals and the hierarchy of their importance, as recognised by the management; it must also take into account the objective, from the point of view of which, an assessment is made. In the literature on the subject, it is argued that efficiency is often equated with effectiveness. The concept of efficiency generally refers to the rule of rational management, formulated in two variants: efficiency (maximising effect) and savings (minimising expenditure) [7].

Information systems should be treated as investments and should be checked as to whether the solutions implemented really do support organisation processes effectively and bring the economic effects desired, in the form of increased profits and/or reduced costs. The effectiveness of the use of Information and communication technologies (ICT) is determined by a number of internal and external factors. Internal factors should, particularly, be taken into account. A detailed analysis of the internal stream is crucial in optimising the efficiency of an enterprise. In order to obtain optimal results from the implementation of a given technology, a model should be built that would be the best for a given enterprise under its conditions. The final condition, therefore, for the use of ICT is a company's own organisational effectiveness [8].

In the authors' research it was assumed that the effectiveness of the implementation of the MES, is understood as the degree of the utilisation of available system functionalities in manufacturing companies on three levels of management, namely, strategic, tactical and operational, as exemplified by a maintenance department. The assessment of the effectiveness of the implementation of the MES, in a production enterprise, is a response to the needs of management boards, regarding the forecasting of the effects of the implementation of new technologies and IT solutions within a company. The research was carried out in two manufacturing companies in mainte-nance departments, in the automotive industry, in which employees work with the help of the MES system, the results of these tests being the basis for the development of a model for evaluating the effectiveness of implementing the MES system within a maintenance department of a production company.

3   Research Results

An analysis of the effectiveness of the use of the MES at the strategic, tactical and operational levels in the maintenance department was carried out in two, automotive industry companies. Production in both enterprises is carried out using a two-shift system. In the first case, the head of the maintenance department supervises the work of 13 employees who service 380 machines, while in the second case, the manager supervises the work of 4 employees who service 20 machines in the company.

Both enterprises manufacture car parts and accessories. In both companies, pro-duction is partly automated. However, in one enterprise, the degree of automation appears somewhat haphazard to a very small extent; in the other enterprise, however, there are larger sections of process automation, in that the enterprise uses professional water cutting systems, CNC machines and industrial robots (Fig. 2).

Fig. 2. An example of the industrial robots within a analysed manufacturing company, source: company's documentation.

The implementation and use of the information system, or the Manufacturing Execution System (MES), may enable further activities in the company, related to the automation of production and may constitute the first element of development, within an enterprise, in line with the Industry 4.0 Concept. In order to analyse the use of the MES at the strategic, tactical and operational levels, in the maintenance departments of Polish manufacturing companies within the automotive sector of industry, we distin-guish the activities, that have been completed: without the MES, partly using the MES and completely using the MES system (Figs. 3, 4, 5 and 6).

The research was carried out in two, automotive industry manufacturing compa-nies, one of which is perceived to be partially automated, while the other is considered

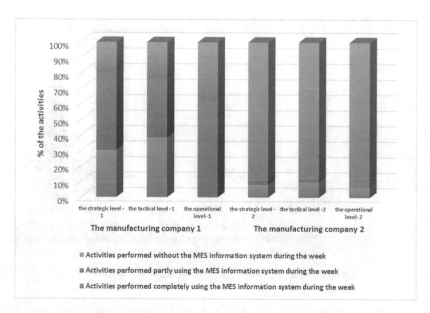

Fig. 3. Activities performed using MES in analysed two manufacturing companies, in the maintenance department

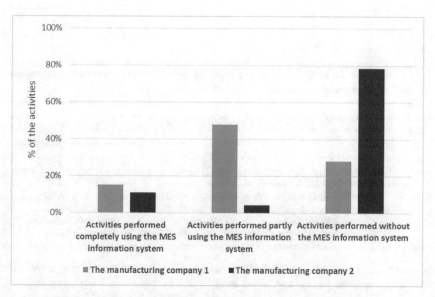

Fig. 4. Activities performed using MES in analysed two manufacturing companies, in the maintenance department at the strategic level.

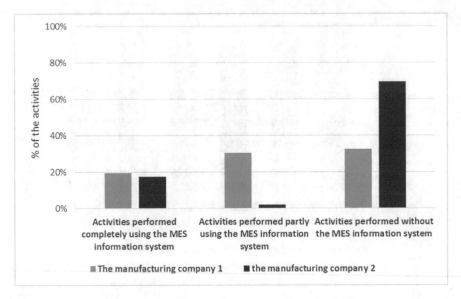

Fig. 5. Activities performed using MES in analysed two manufacturing companies, in the maintenance department at the tactical level.

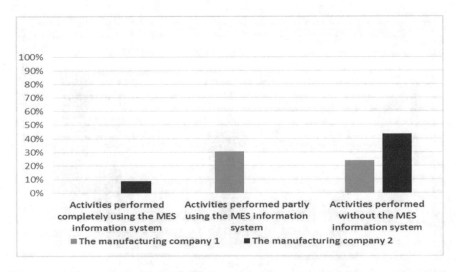

Fig. 6. Activities performed using MES in analysed two manufacturing companies, in the maintenance department at the operational level.

to have a minimum degree of process automation, in the maintenance department, and is supported by the MES. It was found that the activities carried out by the maintenance services during a week's working time were, in both cases, entirely/totally supported by the IT system but only to a very small extent. In one enterprise, the system supports them partially, while in the other, most of the work, carried out by these services, is carried out manually (Fig. 3).

Work done by the employees at the strategic level is also totally automated, to a very small extent. In one of the enterprises, the employees are partially supported by the system to about 50%, while in the second, when it comes to partial system support, this is only about 4%, with total support, however, in this case, amounting to about 11% of all activities. As much as 80% of the activities, performed by these services, in this enterprise, is carried out manually (Fig. 4).

As far as the tactical level is concerned, in both companies, process automation totally supported by the system, is at a similar level (about 20%), but the activities supported partly by the IT system differ to a very large extent; in one enterprise it is about 30% while in the other, it is only 2%. A larger difference occurs in the activities performed manually; in one company, the level is about 33%, while in the other, it is as much as 70% of activities (Fig. 5).

In one enterprise, the operations are totally supported at a level of 9%, while the second enterprise does not have complete, automated support at this level; however, it does have the partial support of the IT system at a level of 30%; in the second case, however, such partial support is missing. On the other hand, manual operations are at a level of 24% and 44% for both enterprises (Fig. 6).

Based on the results of research- and knowing that the maintenance services in one enterprise take care of 380 machines and devices - and in the other enterprise take care of 20 machines/devices- it was found that work in the maintenance department is marginally supported by the IT system MES. These results indicate that the enterprises surveyed do not function under the Industry Concept 4.0, which can be understood as a 'common term combining the technology and organisation of the value-added chain' [9] which assumes the existence of intelligent systems that are networked, that is, vertically connected to other processes within the enterprise and horizontally associated with value creating networks. They can be managed in real time from the moment that an order is placed, up to the co-ordination of sales logistics. These are complex solutions created at the interface between engineering, information technology and management knowledge.

4   Conclusions

Polish manufacturing enterprises, striving to adapt their activities to the requirements of the Industry 4.0 Concept, should implement and use IT systems. Based on the research results it can be stated, however, that in the enterprises surveyed, the manner in which the Industry 4.0 Concept is applied is well known and the need to implement its elements is recognised, especially in enterprises with a significant degree of process automation where the activities, performed by the employees, are partly supported by the IT system. The work related to the Industry 4.0 Concept must focus on cybernetic systems and also on new innovative solutions, both in terms of the products manu-factured and in the methods by which they are produced, in all areas contributing to their production.

Based on the research carried out, it can be observed that in enterprises, there are different levels of intelligent machines and technologies; this has an impact and will have an impact in future, on production areas aimed at the Industry 4.0 Concept; furthermore, it can be said that they will develop at different rates of intensity.

Further directions of research work will include the extension and inclusion of other, automotive industry enterprises, this being the branch of industry, in the field of industrial processing, which qualifies under section C, of the Polish Classification of Activities (PKD 2007) since it has a much larger number of machines/devices, combined with a greater automation of production processes and thus, will generate further possibilities for the comparison of results.

In consequence- and based on the results of the research- a model will be built to assess the effectiveness of the implementation of the MES system in manufacturing enterprises using neural networks.

Acknowledgments. This work is supported by program of the Polish Minister of Science and Higher Education under the name "Regional Initiative of Excellence" in 2019–2022, project no. 003/RID/2018/19, funding amount 11 936 596.10 PLN.

References

1. Lee, J.: Big data environment Industry 4.0. Harting Mag. **26**, 8–10 (2013)
2. Kagermann, H.: Recommendations for implementing the strategic initiative Industrie 4.0. In: Final report of the Industrie 4.0 Working Group, pp. 15–20 (2013)
3. Bocewicz, G., Nielsen, I., Banaszak, Z.: Production flows scheduling subject to fuzzy processing time constraints. Int. J. Comput. Integr. Manuf. **29**(10), 1105–1127 (2016)
4. Dudycz, T., Brycz, B.: Effectiveness of Polish enterprises in the years 1994–2004 - preliminary empirical research. Scientific Papers of the Social Academy of Entrepreneurship and Management, Poland (2006)
5. Remenyi, D., Money, A., Bannister, F.: The Effective Measurement and Management of ICT Cost and Benefits. Elsevier, Oxford (2007)
6. Simon, H.: Making Decisions and Managing People in Business and Administration. Helion, Poland (2007)
7. Matwiejczuk, R.: Efficiency - an attempt to interpret, Organization Overview 11 (2000)
8. Behr, K.: The value effectiveness, efficiency, and security of IT controls. In: An Empirical Analysis. https://www.academia.edu/Documents/in/IT_and_Operations. Accessed 30 Jan 2018
9. Hermann, M., Pentek, T., Otto, B.: Design principles for Industrie 4.0 scenarios. In: A Literature Review, Working Paper, Technische Universität Dortmund (2015)

Using the Simulation Method for Modelling a Manufacturing System of Predictive Maintenance

Sławomir Kłos$^{(\boxtimes)}$ ⓘ and Justyna Patalas-Maliszewska ⓘ

Faculty of Mechanical Engineering, University of Zielona Góra,
Licealna 9, 65-417 Zielona Góra, Poland
{s.klos, j.patalas}@iizp.uz.zgora.pl

Abstract. The Industry 4.0 concept assumes the implementation of predictive maintenance as an integral part of manufacturing systems. The parallel-serial manufacturing system includes groups of redundant resources. In the case of damage or malfunction, one or other of them could complete manufacturing operations. In this paper, an analysis of the performance and average lifespan of the products of the parallel-serial manufacturing system is presented, for different methods of material flow control. The parallel-serial manufacturing system is considered where the availability of resources and buffer capacity is the input value and the throughput and average lifespan of the products, that is, the time that their details remain in the system, is the output value. The performance of the system is analysed, using different dispatching rules which are allocated to the manufacturing resources. The simulation model of the system is created using Tecnomatix Plant Simulation.

Keywords: Predictive maintenance · Computer simulation ·
Parallel-Serial manufacturing system · Throughput ·
Average lifespan of product

1 Introduction

1.1 Predictive Maintenance

Predictive maintenance techniques enable the conditions of a service manufacturing system to be determined in order to estimate when and where maintenance activity is required. The Industry 4.0 concept assumes that smart manufacturing factories should be resistant to those failures which result when creating self-organising manufacturing systems with redundant resources. The parallel-serial manufacturing systems includes redundant resources that are connected in such a manner as to guarantee the undisturbed flow of material even in the case of partly - or fully - impaired manufacturing resources. The structure of the parallel-serial manufacturing system is resistant to

This work is supported by program of the Polish Minister of Science and Higher Education under the name "Regional Initiative of Excellence" in 2019–2022, project no. 003/RID/2018/19, funding amount 11 936 596.10 PLN.

resource failure, with the proviso that each resource is redundant. In this paper, control of the flow of material is based on the dispatching rules allocated to the manufacturing resources and buffers. The dispatching rules decide which successor the material will be sent to. The simulation model of the system is prepared using Tecnomatix Plant Simulation Software. On the basis of prepared simulation experiments, the impact of the dispatching rules, the capacity of the buffers and the availability of resources on the throughput and average lifespan of the system, is analysed. The research problem can be formulated as follows: *Given, is a simulation model of a parallel-serial manufacturing system. What is the impact on the effectiveness of the manufacturing system, of the capacity of the buffers, the availability of resources and the dispatching rules which are allocated to the resources?*

1.2   Literature Overview

Many authors have studied the optimal predictive maintenance strategies in manufacturing enterprises [1–3]. Renna is engaged in the evaluation of manufacturing-system performance in dynamic conditions, when various maintenance policies are being implemented in a multi-machine manufacturing system, controlled by multi-agent-architecture. A discrete simulation environment has been developed in order to investigate the performance measures and the indices of the costs of maintenance policies [4]. Mokhtari et al. suggest joint production and maintenance scheduling (JPMS), with multiple preventive maintenance services, in which the reliability/availability approach is employed to model the maintenance aspects of a problem. They propose developing a mixed-integer nonlinear-programming model and a population-based variable-neighbourhood-search algorithm, as a solution method [5]. Boschian et al. compare two strategies for operating a production system composed of two machines working in parallel and a downstream inventory supplying an assembly line; manufacturing resources, which are prone to random failures, undergo preventive and corrective maintenance operations. A simulation model for each strategy is developed so as to be able to compare them and to simultaneously determine the timing of preventive maintenance on each machine, considering the total average cost per time unit as the performance criterion [6]. Wan et al. analysed the issue of active preventive maintenance and the related system architecture for manufacturing big data solutions in smart manufacturing systems [7]. The data collection and data processing stages with respect to the Cloud computing environment were presented. Ni and Jin presented new decision support tools based on mathematical algorithms and simulation tools for effective maintenance operations [8]. The system enables the short-term throughput of bottleneck identification, the estimation of maintenance windows of opportunity, the prioritization of maintenance tasks, the joint production and maintenance scheduling systems and maintenance staff management. The system was implemented in automotive manufacturing area. Kłos and Patalas-Maliszewska propose the implementation of simulation tools for the predictive maintenance of manufacturing systems [9, 10]. They analysed the impact of the allocation of buffers and the topology of the manufacturing system on the throughput and average lifespan of the products. Different structures of parallel-serial manufacturing systems were taken into account. The proposed model for the intelligent maintenance management system (IMMS), is based on the results of simulation experiments which were conducted using Tecnomatix Plant Simulation software.

2 Simulation Model of a Manufacturing System

The parallel-serial manufacturing system includes three production lines with each line realising the same two technological operations (see Fig. 1). The proposed structure guarantees deadlock and a starvation-free system. The same structures could be used for example in multimodal transportation networks [11].

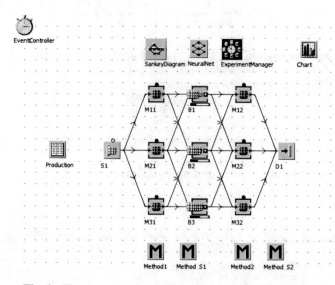

Fig. 1. The structure of a parallel-serial manufacturing system

The machine groups (M_{11}, M_{21}, M_{31}) and (M_{12}, M_{22}, M_{32}) are similar and realise the same technological operations but the availability of the machines is different, depending on the exploitation period. Between each pair of machines, buffers are allocated where parts are stored after machining. The parts are sent from the machines to buffers, according to dispatching rules. Three dispatching rules implemented in the Tecnomatix Plant Simulation software are taken into account:

1. Cyclic.
2. Minimum contents.

For example, machine M_{21} is connected to buffers B_1, B_2 and B_3. The cyclic dispatching rules send the parts, in sequence, to the three buffers. If, however, any buffer is full, the rules state that the part should be sent to the next buffer and if the option 'Block' has been activated, then the machine must wait until the buffer has free space. The minimal contents dispatching rule decides to send the part into the buffer with the most space while the minimal relative occupancy dispatching rule sends the part to the buffer with statistical minimal occupancy. In the system, four kinds of product are manufactured A, B, C, D. Each product is manufactured in two, different lot sizes. The changing of the lot size requires additional setup time on the machines. The dispatching rules are allocated only on machines: M_{11}, M_{21}, M_{31}. The processing and set-up times of the manufacturing resources are determined on the basis of a lognormal distribution. A lognormal distribution is a continuous distribution in which a random number has a natural logarithm corresponding to a normal distribution. The realisations are non-negative, real numbers. The density of the lognormal distribution Lognor (σ, μ) is calculated as follows:

$$f(x) = \frac{1}{\sigma_0 x \sqrt{2\pi}} \cdot \exp\left[\frac{-\ln(x - \mu_0)^2}{2\sigma_0^2}\right] \tag{1}$$

where σ and μ are respectively mean and standard deviations and are defined as follows:

$$\mu = \exp\left[\mu_0 + \frac{\sigma_0}{2}\right] \tag{2}$$

$$\sigma^2 = \exp(2\mu_0 + \sigma_0^2) \cdot (\exp(\sigma_0^2) - 1) \tag{3}$$

The maximum of the density function is defined as:

$$\exp(\mu_0 - \sigma_0^2) \tag{4}$$

The values of operation times for all machines are defined differently for each product:

- product A - $\sigma^2 = 480$ and $\mu = 20$,
- product B - $\sigma^2 = 580$ and $\mu = 40$,
- product C - $\sigma^2 = 180$ and $\mu = 20$,
- product D - $\sigma^2 = 280$ and $\mu = 20$.

The values of setup times are defined as $\sigma^2 = 2400$ and $\mu = 500$. In the system, four products (A, B, C, D) are manufactured, based on the following sequence of the size of the production batches: A - 100, B - 300, C - 80, D - 120, A - 50, B - 140, C - 220, D - 60.

As input values for the simulation experiments, the following variables are taken into consideration: the allocation of buffer capacity and the availability of manufacturing resources, while for the output values, the variables are: the throughput per hour and the average lifespan of the products (that is, the average time that the part spends in the system). It was assumed that the availability of manufacturing resources could be changed from 90% to 100% and that the buffer capacity could be changed from 1 to 15. In Table 1, the input data for 50 simulation experiments is presented with the allocation of buffer capacity and machine availability, expressed as a percentage. The data is prepared manually, using a random generator. Figure 2, shows the total buffer capacity, that is, the sum of the buffer capacities, for each experiment. The maximum, buffer capacity value is 45 (Exp 39) where each buffer has a capacity of 15. The smallest buffer capacity is 1 and is used in experiments Exp 07, Exp 08, Exp 09 and Exp 19. The maximum, average availability of resources is 100%, since all resources are failure free, while the smallest is 92% (Exp 01). The simulation experiments are conducted for two dispatching rules, viz. the Cyclic and the Minimum contents and feature the 'Blocked' and 'Not blocked' options.

Table 1. Input data for 50 simulation experiments

Exp	B_1	B_2	B_3	M_{11}	M_{21}	M_{31}	M_{12}	M_{22}	M_{32}	Exp	B_1	B_2	B_3	M_{11}	M_{21}	M_{31}	M_{12}	M_{22}	M_{32}
01	1	3	1	92	92	91	91	95	93	26	3	3	3	100	100	100	100	100	100
02	1	2	2	90	100	91	94	93	97	27	5	5	5	100	100	100	100	100	90
03	1	4	3	90	95	92	95	96	94	28	5	5	5	90	100	100	100	100	100
04	1	3	2	95	98	92	91	98	93	29	5	5	5	100	90	100	100	100	100
05	1	2	4	96	91	95	92	97	98	30	10	10	10	100	90	100	100	90	100
06	5	2	1	98	91	92	93	96	94	31	5	5	5	100	100	90	100	100	100
07	1	1	1	100	100	100	100	90	100	32	10	10	10	90	100	100	90	100	100
08	1	1	1	100	100	100	100	100	90	33	10	10	10	100	100	90	100	100	90
09	1	1	1	100	90	100	100	100	100	34	5	5	5	98	98	98	100	97	100
10	10	10	10	100	100	100	90	90	90	35	4	4	4	100	100	100	100	100	100
11	5	3	3	98	90	90	96	96	97	36	5	5	5	100	100	100	100	100	100
12	3	2	5	94	96	98	92	96	96	37	5	5	5	100	100	100	100	100	100
13	1	4	3	99	96	100	94	93	98	38	5	5	5	100	100	100	100	100	100
14	4	5	2	95	94	93	97	95	99	39	15	15	15	100	100	100	100	100	100
15	4	3	3	99	93	92	97	97	93	40	10	10	10	100	100	100	100	100	100
16	10	10	10	90	90	90	100	100	100	41	10	10	10	90	100	100	100	100	100
17	5	1	5	92	95	93	98	99	97	42	10	10	10	100	90	100	100	100	100
18	2	5	1	98	94	97	99	97	96	43	10	10	10	100	100	90	100	100	100
19	1	1	1	100	100	100	100	100	100	44	10	10	10	100	100	90	90	100	100
20	4	2	4	94	97	99	94	97	98	45	10	10	10	100	100	100	90	100	100
21	2	2	2	100	100	100	100	100	100	46	10	10	10	100	100	100	100	90	100
22	10	10	10	90	100	90	100	90	100	47	10	10	10	100	100	100	100	100	90
23	10	10	10	90	90	100	100	100	100	48	10	10	10	90	100	90	100	100	90
24	5	5	5	100	100	100	90	100	100	49	10	10	10	100	90	100	90	100	90
25	5	5	5	100	100	100	100	90	100	50	10	10	10	90	90	100	100	90	90

3 The Results of the Simulation Experiments

The simulation experiments are realised using Tecnomatix Plant Simulation, v 12.2. For each experiment, three observation were made. The time of each experiment is 80 real hours (10 days). The results of the simulation experiments are presented in Figs. 2, 3, 4 and 5. Generally speaking, the best throughput results were achieved using the Minimum contents dispatching rule where the throughput was 23, 14 products per hour (Exp 39). The poorest average lifespan value was similar for both dispatching rules in Exp 10 and was 1:30:54. In Exp 10, machines M_{11}, M_{21}, M_{31} had maximum, 100% availability while machines M_{12}, M_{22}, M_{32} had 90% availability, that is, were 'bottlenecked'. The buffers had ample space, relatively speaking since each buffer has a capacity of 10 and the configuration of the results of the manufacturing system indicated a high, average lifespan value, suggesting a high level of work-in-progress. Where the Block option is activated, the throughput value could be some 3%–6% lower than where the option has not been activated. This is especially the case with the Cyclic dispatching rule.

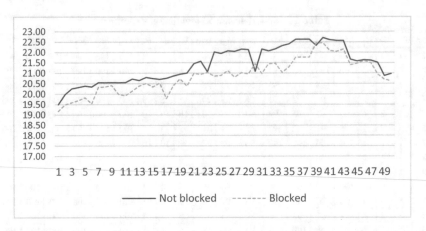

Fig. 2. The results of simulation experiments – throughput per hour, for the cyclic dispatching rule

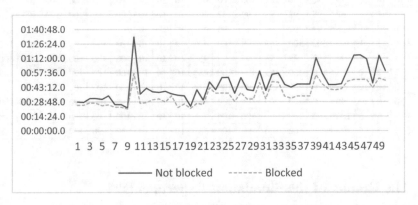

Fig. 3. The results of simulation experiments – average lifespan for the Cyclic dispatching rule

The smallest average lifespan level is reached for experiments with single capacity buffers, but in these cases, manufacturing systems are sensitive to failure and throughput is relatively small. The impact of the Block function on the throughput and average lifespan is important for the Cyclic dispatching rule. The average availability of the machines, within the system is in the range 92%–100%. The system is the least sensitive to failures in Exp 50, this being the highest value of throughput for this level of availability.

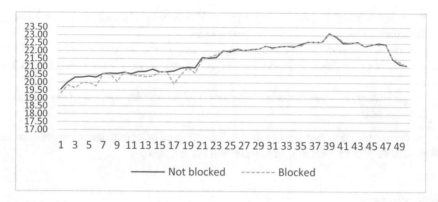

Fig. 4. The results of simulation experiments - throughput per hour for the Minimum contents dispatching rule

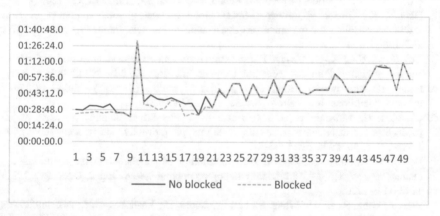

Fig. 5. The results of simulation experiments - average lifespan for the Minimum contents dispatching rule

4 Conclusions

In the paper, the impact of the dispatching rules on the effectiveness of a parallel-serial manufacturing system using simulation methods is analysed. 50 simulation experiments was conducted for different allocations of buffer capacities and availabilities of manufacturing resources. The results of the experiments enable the following conclusions to be formulated:

- allocation of the Minimum contents dispatching rule gives higher throughput values and the same values of average lifespan than does the Cyclic dispatching rule,
- the highest throughput values are reached with maximum, buffer capacity and 100% availability of resources,
- one system configuration, namely the allocation of buffers and availabilities, resulted in an extremely high, average product lifespan; this can be seen in Exp 10,

- the parallel serial manufacturing system is relatively resistant to failure; system throughput changes are in the range 1%–15%.
- for low availability of manufacturing resources, the buffer capacity should be greater, in order that a high, system throughput value, may be obtained (Exp 01 vs Exp 50),
- the parallel-serial manufacturing systems could realise the concept of the predictive maintenance.

In further research, the impact of the competences of maintenance workers on the effectiveness of a manufacturing system will be analysed.

References

1. Driessen, J.P.C., Peng, H., Houtum, G.J.: Maintenance optimisation under non-constant probabilities of imperfect inspections. Reliab. Eng. Syst. Saf. **165**, 115–123 (2017)
2. Dhouib, K., Gharbi, A., Aziza, M.N.B.: Joint optimal production control/preventive maintenance policy for imperfect process manufacturing cell. Int. J. Prod. Econ. **137**(1), 126–136 (2012)
3. Yeh, R.H., Kao, K.-C., Chang, W.L.: Optimal preventive maintenance policy for leased equipment using failure rate reduction. Comput. Ind. Eng. **57**, 304–309 (2009)
4. Renna, P.: Influence of maintenance policies on multi-stage manufacturing systems in dynamic conditions. Int. J. Prod. Res. **50**, 345–357 (2012)
5. Mokhtari, H., Mozdgir, A., Kamal, Abadi I.: A reliability/availability approach to joint production and maintenance scheduling with multiple preventive maintenance services. Int. J. Prod. Res. **50**, 5906–5925 (2012)
6. Boschian, V., Rezg, N., Chelbi, A.: Contribution of simulation to the optimization of maintenance strategies for a randomly failing production system. Eur. J. Oper. Res. **197**(3), 1142–1149 (2009)
7. Wan, J., Tang, S., Li, D., Wang, S., Liu, C., Abbas, H., Vasilakos, A.V.: A manufacturing big data solution for active preventive maintenance. IEEE Trans. Ind. Inform. **13**, 2039–2047 (2017)
8. Ni, J., Jin, X.: Decision support systems for effective maintenance, operations. CIRP Ann. Manufact. Technol. **61**, 411–414 (2012)
9. Kłos, S., Patalas-Maliszewska, J.: The use of the simulation method in analysing the performance of a predictive maintenance system. In: 15th International Conference on Distributed Computing And Artificial Intelligence, Special Sessions, Advances in Intelligent Systems and Computing, vol. 801, pp. 42–49. Springer Nature Switzerland (2019)
10. Kłos, S., Patalas-Maliszewska, J.: Using a simulation method for intelligent maintenance management. In: The First International Conference on Intelligent Systems in Production Engineering and Maintenance - ISPEM 2017, Advances in Intelligent Systems and Computing, vol. 637, pp. 85–95. Springer International Publishing (2018)
11. Bocewicz, G.: Robustness of multimodal transportation networks. Eksploatacja i Nieza-wodność-Maintenance and Reliability **16**(2), 259–269 (2014)

UAV Mission Planning Subject to Weather Forecast Constraints

A. Thibbotuwawa[1], G. Bocewicz[2(✉)], P. Nielsen[1], and Z. Banaszak[2]

[1] Department of Materials and Production,
Aalborg University, Aalborg, Denmark
{amila,peter}@mp.aau.dk
[2] Department of Computer Science and Management,
Koszalin University of Technology, Koszalin, Poland
bocewicz@ie.tu.koszalin.pl,
Zbigniew.Banaszak@tu.koszalin.pl

Abstract. A multi-trip UAV delivery problem is considered in which trajectories are planned for UAVs operating in a hostile environment. UAV battery capacity and payload weight as well as vehicle reuse are taken into account. A fleet of homogeneous UAVs fly in a 2D plane matching a distribution network to service customers in a collision-free manner. The goal is to obtain a sequence of sub-missions that will ensure delivery of requested amounts of goods to customers, satisfying their demands within a given time horizon under the given weather forecast constraints. In this context, our objective is to establish the relationships linking decision variables such as wind speed and direction, battery capacity and payload weight. Computational experiments which allow to assess alternative strategies of UAV sub-mission planning are presented.

Keywords: UAV routing · UAV fleet mission planning · Delivery service

1 Introduction

As energy consumption in unmanned aerial vehicles (UAVs) mostly depends on their payload and the speed and direction of wind, UAV mission planning strategies have to estimate energy consumption based on these factors in order to identify the set of all reachable destinations. Due to changing weather conditions, large numbers of customers serviced and requested deliveries, quite long distances between delivery points and a relatively long time horizon of supply services, a UAV fleet mission should be planned as a sequence of repeatedly executed sub-missions to ensure step-by-step delivery to customers to satisfy their demands within a given time horizon. In this context, the present study addresses the issue of sub-mission planning problems, i.e. UAV fleet routing and scheduling problems. Such problems must take into account the changing weather conditions, relationships among decision variables such as wind speed and direction, battery capacity and payload weight, and sub-mission feasibility, i.e. whether a given level of customer satisfaction is reachable within the sub-mission's time window. Similar problems have been considered in [6, 7].

© Springer Nature Switzerland AG 2020
E. Herrera-Viedma et al. (Eds.): DCAI 2019, AISC 1004, pp. 65–76, 2020.
https://doi.org/10.1007/978-3-030-23946-6_8

In this connection, the focus of the present study is on solutions that allows one to find an admissible (collision-free) plan of delivery sub-missions composed of a sequence of UAV multi-trip-like flights maximizing order fulfillment. Therefore, the reference model of a UAV-driven distribution system considered here takes into account data related to the weather forecast, number of customers, customer demand, fleet size, parameters describing a fleet of homogeneous UAVs such as payload and battery capacity, flight distances among customers and depots, the time horizon and so on. When the weather forecast is known, the time horizon can be subdivided into so-called weather time windows (with the same wind speed and direction), which in turn can be arbitrarily subdivided into so-called flying time windows (i.e., periods during which a UAV of a given energy limit can fly). The number of flying time windows generates the number of delivery sub-missions. For each sub-mission, a UAV fleet mission plan subject to weather forecast constraints is considered. Of course, in each such case, it is necessary to analyze different factors including, e.g. payload weight, which influence energy consumption, the number or size of payloads that a UAV can accommodate as well as the number of depots, and flight distances, all of which constrain the quantity of products that can be delivered. Our study presents a declarative framework which allows to formulate a reference model for the analysis of the relationships between the structure of a given UAV-driven supply network and its potential behavior, resulting in a sequence of sub-missions following a required delivery. The reported computational experiments provide requirements for a solvable class of UAV-driven mission planning problems subject to weather forecast and energy consumption constraints. The results fall within the scope of research previously reported in [3] and [4]. The remainder of the article is structured as follows: Sect. 2 provides an overview of literature. A reference model for a UAV fleet routing and scheduling problem is discussed in Sect. 3. The problem is formulated in Sect. 4, which also presents a Constraint Optimization Problem-based method for planning UAV delivery missions. An example illustrating the approach used is given in Sect. 5. Conclusions are formulated and directions of future research are suggested in Sect. 6.

2   Literature Review

UAV mission planning problems can involve a significant amount of uncertain information, in contrast to vehicle routing problems (VRP), because the former type of vehicles have the ability to change, modify, and optimize their routes in 3D space [5, 12]. The task of mission planning is to find a sequence of waypoints that connect the start to the destination location; this differs from trajectory planning, in which the solution path is expressed in terms of the degrees of freedom of the vehicle [10, 13].

Some of the existing models have placed less focus on (most of) the physical properties related to UAVs and have used approaches such as VRP with time windows (VRPTW), which include a large number of targets and vehicles [1, 9, 14, 17]. Some studies have proposed to cluster the network to reduce problem size, taking into account the relative capabilities of UAVs [11, 15]. This study proposes to cluster customer nodes, given that a set of feasible UAV fleet routings with schedules can be derived for each cluster subject to weather-dependent energy consumption constraints.

Various technical and environmental factors influence the potential search for possible UAV mission planning solutions. These factors encompass the technical parameters of UAVs (UAV dimensions, battery capacity and carrying payload limit) and the environmental aspects of the changing weather conditions, including wind speed, wind direction and air density. Existing research has largely ignored the impact of weather conditions by using linear approximations for energy consumption [6]. There do, however, exist studies which consider the influence of wind conditions on energy consumption in UAV mission planning [16]. As the linear approximations reported in current literature cannot be used in this study due to the payload weights considered [6], non-linear models proposed in previous work are used to calculate energy consumption in relation to weather conditions [18]. More complexity to finding a solution is added by the problem of avoidance of collisions against fixed and flying objects [8]. Collision avoidance can be achieved by predicting potential collisions in offline planning or by reacting to collisions registered by sensors in online planning [2, 8]. It must also be remembered that different approaches to avoiding collisions are needed for a fleet of UAVs flying in unrestricted space vs. UAVs fly in dedicated corridors.

There is little knowledge on how the above-mentioned parameters can affect solutions to UAV mission planning problems, even with reference to deterministic approaches. The existing literature also fails to provide solutions that would accommodate the effects of changing weather conditions on energy consumption in UAVs.

3 Modeling

3.1 Assumptions

The idea of the considered problem is presented in Fig. 1. Given is a set of customers located at different points of a delivery distribution network. These customers are to be serviced by a fleet of UAVs during a specified time horizon, under changing weather. In this context, the following assumptions are taken into account:

- The weather forecast is known in advance with sufficient accuracy to specify the so-called weather time windows W_l.
- The weather time windows can be subdivided into flying time windows F_l.
- Wind speed vw_l and the direction of wind θ_l for each F_l are known in advance. Wind direction is the same inside a given weather time window.
- Every route travelled starts and terminates within a given flying time window.
- All UAVs are charged to their full energy capacity before the start of a flying time window, and one UAV can only fly one time during a flying time window.
- The same kind of cargo is delivered to different customers in different amounts [kg].
- Network G consists of customer location points and flying corridors.

The goal is to fulfill all customer demands, such that each customer is at required service level before the end of the time horizon and all constraints related to energy limits and congestion avoidance are satisfied. The proposed approach assumes that the process of finding solutions takes place at two levels: Mission and Sub-mission Planning (see Fig. 1). At the first level, the distribution network G is divided into a set

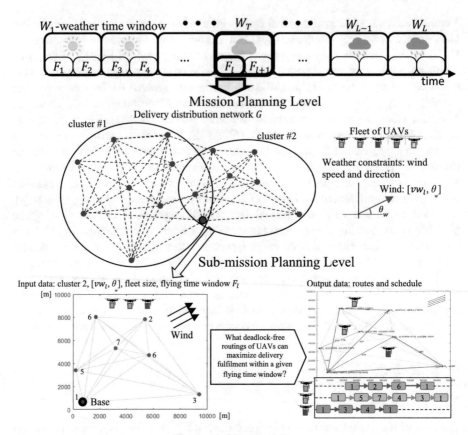

Fig. 1. Problem modeling

of clusters (covering the base and several customers) for which the size of the UAV fleet is determined. At the second level, UAV sub-missions (specified by UAV routes and schedules) maximizing customer satisfaction are calculated for each cluster. It is assumed that there exists a sequence of sub-missions which allow to fulfill all customer demands within the given time horizon.

3.2   Declarative Model

The mathematical formulation of the model considered employs the following parameters, variables, sets and constraints:

Parameters

Network

$G = (N, E)$ graph of a transportation network: $N = \{1 \ldots n\}$ is a set of nodes, $E = \{\{i, j\} | i, j \in N, i \neq j\}$ is a set of edges

$CL_{m,l} = (N_{m,l}, E_{m,l})$ subgraph of G representing the mth cluster in the lth flying time window: $N_{m,l} \subseteq N$ and $E_{m,l} \subseteq E$

z_i demand at node $i \in N$, $z_1 = 0$

$d_{i,j}$	travel distance from node i to j
$t_{i,j}$	travel time from node i to j
w	time spent on take-off and landing of a UAV
$b_{\{i,j\};\{\alpha,\beta\}}$	binary variable corresponding to crossed edges.

$$b_{\{i,j\};\{\alpha,\beta\}} = \begin{cases} 1 & \text{when edges} \{i,j\} \text{ and } \{\alpha, \beta\} \text{ are crossed} \\ 0 & \text{otherwise.} \end{cases}$$

UAV Technical Parameters

K	size of the fleet of UAVs
Q	maximum loading capacity of a UAV
$vg_{i,j}$	ground speed of a UAV travelling from node i to node j
C_D	aerodynamic drag coefficient of a UAV
A	front facing area of a UAV
ep	empty weight of a UAV
D	air density
b	width of a UAV
E_{max}	maximum energy capacity of a UAV

Environmental Parameters

H	time horizon $H = [0, t_{max}]$
W_T	weather time window T: $W_T = [WS_T, WE_T]$, WS_T/WE_T is a start/end time of W_T
F_l	flying time window l: $F_l = [FS_l, FE_l]$, FS_l/FE_l is a start time of F_l
vw_l	wind speed in the lth flying time window
θ_l	wind direction in the lth flying time window
$va^l_{i,j}$	airspeed of a UAV traveling from node i to node j in the lth flying time window,
$P^k_{i,j}$	energy consumed by the kth UAV per one time unit, during the flight from node i to j,

Decision Variables

$x^k_{i,j}$ binary variable used to indicate if the kth UAV travels from node i to node j

$$x^k_{i,j} = \begin{cases} 1 & \text{if } k^{th} \text{ UAV travels from node i to node j} \\ 0 & \text{otherwise.} \end{cases}$$

y^k_i	time at which the kth UAV arrives at node i
c^k_i	payload weight delivered to node i by the kth UAV
$f^k_{i,j}$	payload weight carried by a UAV from node i to j by the kth UAV
$\pi^k_{m,l}$	route of the kth UAV in the mth cluster in the lth flying time window

$$\pi_{m,l}^k = (v_1, \ldots, v_i, v_{i+1}, \ldots, v_\mu), v_i \in N_{m,l}, x_{v_i,v_{i+1}}^k = 1$$

Sets

Y^k set of times y_i^k – schedule of the kth UAV

Y family of Y^k – schedule of UAV fleet

C^k set of c_i^k – payload weight delivered by the kth UAV

C family of C^k

Π set of UAV routes $\pi_{m,l}^k$

$S_{m,l}$ Sub-mission in the mth cluster in the lth flying time window $S_{m,l} = (\Pi, Y, C)$

Constraints

Arrival Time at Nodes. Relationship between the binary decision variable of $x_{i,j}^k$ and the variable of y_i^k.

$$(x_{i,j}^k = 1) \Rightarrow (y_j^k = y_i^k + t_{i,j} + w), \forall \{i,j\} \in E_{m,l}, k = 1 \ldots K \tag{1}$$

Collision Avoidance. The crossed edges $(b_{\{i,j\};\{\alpha,\beta\}} = 1)$ cannot be used at the same time when they are occupied by the UAVs ($x_{i,j}^k = 1$ and $x_{\alpha,\beta}^v = 1$):

$$\left[\left(b_{\{i,j\};\{\alpha,\beta\}} = 1 \right) \wedge \left(x_{i,j}^k = 1 \right) \wedge \left(x_{\alpha,\beta}^v = 1 \right) \right] \Rightarrow \left[(y_j^k \leq y_\alpha^v) \vee (y_i^k \leq y_\beta^v) \right] \tag{2}$$
$$k, v = 1 \ldots K, k \neq v \text{ and } \{i,j\}, \{\alpha, \beta\} \in E_{m,l}$$

Capacity. The demand assigned to a UAV should not exceed its capacity.

$$c_i^k \geq 0, \forall i \in N_{m,l}, \tag{3}$$

$$c_j^k \leq Q \times \sum_{i \in N_{m,l}} x_{i,j}^k, \forall j \in N_{m,l}, k = 1 \ldots K, \tag{4}$$

$$\sum_{i \in N_{m,l}} c_i^k \leqslant Q, \quad k = 1 \ldots K. \tag{5}$$

$$\sum_{k=1..K} c_i^k \leqslant z_i, \forall i \in N_{m,l}, \tag{6}$$

The sum of all weights c_j^k carried by an UAV should not exceed the maximum carrying payload Q and demand at node z_j.

Flow of UAVs. When a UAV arrives at a node, that UAV must leave from that node.

$$\sum_{j \in N_{m,l}} x_{1,j}^k = 1, k = 1 \ldots K, \tag{7}$$

$$\sum_{j \in N_{m,l}} x_{i,j}^k = \sum_{j \in N_{m,l}} x_{j,i}^k, \quad \forall i \in N_{m,l}, k = 1 \ldots K, \tag{8}$$

Energy. A UAV has a maximum energy capacity of E_{max}, which means that when in flight, it cannot consume energy in excess of E_{max}.

$$\sum_{i \in N_{m,l}} \sum_{j \in N_{m,l}} x_{i,j}^k P_{i,j}^k t_{i,j} \leqslant E_{max}, \qquad k = 1 \ldots K \tag{9}$$

$$P_{i,j}^k = \frac{1}{2} C_D A D (va_{i,j}^l)^3 + \frac{\left(ep + f_{i,j}^k\right)^2}{D b^2 va_{i,j}^l}, \tag{10}$$

$$va_{i,j}^l = \sqrt{\left(vg_{i,j} \cos \vartheta_{i,j} - vw_l \cos \theta_l\right)^2 + \left(vg_{i,j} \sin \vartheta_{i,j} - vw_l \sin \theta_l\right)^2}. \tag{11}$$

Cost Function. The cost function is defined as a mean customer satisfaction level:

$$CS = \frac{1}{|N_{m,l}|} \sum_{i \in N_{m,l}} \sum_{k=1..K} c_i^k. \tag{12}$$

4 Problem Formulation

The review of literature shows that there is a scarcity of effective solutions for planning UAV missions under changing weather conditions. To find a solution to this type of problem, one has to answer the following question:

Consider a UAVs fleet of size K servicing, in the lth flying time window (F_l), customers allocated in the mth cluster of the delivery distribution network (subgraph $CL_{m,l}$). Does there exist sub-mission $S_{m,l}$ (determined by variables Π, Y, C) which can maximize customer satisfaction CS under constraints related to energy consumption (9)–(12), collision avoidance (2), etc.)?

The investigated problem can be viewed as a Constraint Optimization Problem (COP) [18] given by (13):

$$COP = (\mathcal{V}, \mathcal{D}, \mathcal{C}, CS), \tag{13}$$

where: $\mathcal{V} = \{\Pi, Y, C\}$ – a set of decision variables including: Π – a set of UAV routes,
Y – a schedule of a UAV fleet, C – a set of payload weights delivered by the UAVs,
$\mathcal{D}$ – a finite set of decision variable domain descriptions,
$\mathcal{C}$ – a set of constraints specifying the relationships between UAV routes, UAV schedules and transported materials (1)–(12),
CS – a cost function representing the level of customer satisfaction.

To solve a constraint optimization problem involves determining such values of the decision variables from the adopted set of domains for which the given constraints are satisfied and the cost function reaches its maximum. Implementation of a COP in a constraint programming environment such as IBM CPLEX enables the construction of a computational engine that can be implemented in an interactive DSS system.

5   Computational Experiments

Consider cluster #2 from Fig. 1 containing 7 nodes (one base and 6 customers). The demands of the nodes are known: z = (0, 60, 70, 30, 40, 20, 30). To service the customers, a fleet of three UAVs is used. The flight parameters which determine the amount of energy consumed (9)–(11) are shown in Table 1.

Table 1. Flight parameters for constraints (9)–(11)

Maximum loading capacity of a UAV	$Q = 80$ kg
Ground speed of a UAV traveling from node i to j	$vg_{i,j} = 20$ m/s
Aerodynamic drag coefficient of a UAV	$C_D = 0.54$
Front facing area of a UAV	$A = 1.2$
Empty weight of a UAV	$ep = 42$ kg
Air density	$D = 1.225$
Width of a UAV	$b = 8.7$
Maximum energy capacity of a UAV	$E_{max} = 6000$ kJ

The goal is to find the sub-missions (routes and schedules) which will maximize customer satisfaction CS (12), assuming that the wind, blowing at 10 m/s, is changing its direction from 30° to 120°. The obtained sub-missions (shown in Fig. 2) provide routes and schedules which guarantee collision-free transport of materials and maximum customer satisfaction $CS = 1$. The results are solutions to problem COP (13) implemented in the IBM CLPEX environment (Windows 10, Intel Core Duo2 3.00 GHz, 4 GB RAM). Note that the different sub-missions all fulfill customer demands but differ in energy consumption (percentage of battery power consumed by each UAV): [85% 87% 96%] – Fig. 2(a); [98% 88% 80%] – Fig. 2(b). The highest energy consumption is observed for the wind direction of 30°.

It is worth noting that optimal solutions obtained under one type of weather conditions may be unattainable under other weather conditions. Alternatively stated, the solutions obtained are characterized by different robustness to changes in weather conditions. Figure 3 shows radar charts illustrating the maximum wind speed (as a function of wind direction) at which the given sub-mission (sub-mission in Fig. 2) guarantees that the given deliveries are all met. For example, the vector of Fig. 3a means that the permissible speed of wind blowing at 60° for the mission of Fig. 2a is 11.1 [m/s].

Radar charts also indicate the maximum wind speed vm at which the given deliveries can be completed regardless of the direction of the wind. The maximum wind speed for the missions of Figs. 3 and 2b is 9.8 and 9.2 [m/s], respectively. It is easy to see that the mission of Fig. 2b is the most robust to changing weather conditions, i.e. the permissible wind speed at which the execution of orders is guaranteed is 9.2 m/s. Finding such solutions is critical for ensuring adequate mission security, i.e. minimizing the risk of failure of the planned mission. In the consideration context, Fig. 4 shows a sub-mission with the highest robustness to changes in wind speed $vm = 11.2$ [m/s].

To evaluate the influence of weather conditions on the obtained solution, experiments for different battery capacities (3000 kJ–6000 kJ), wind speeds (5 m/s–15 m/s) and wind directions (45, 135, 225, 325) were carried out. The results are given in Fig. 5. As one can see, the drop in the value of CS is nonlinear. An increase in wind speed from 10 to 15 m/s results in a greater decrease in CS ($\Delta CS \in [0, 0.5]$) than in the case of the change from 5 to 10 m/s ($\Delta CS \in [0, 0.34]$). This means that a relatively small change in weather conditions can significantly affect the value of the objective function of CS (often reducing the number of optimal solutions to CS = 1).

The experiments of this type indicate that the adopted model (1)–(12) has limitations that restrict its use to mission planning problems in distribution networks in which the number of nodes does not exceed 12 and the number of UAVs in the fleet is not larger than 5. For such problem instances, solutions can be obtained in under 1000 s.

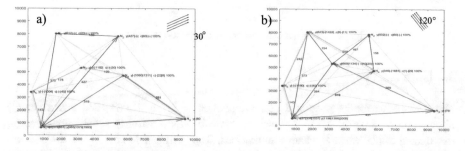

Fig. 2. Solutions obtained: $vw = 10$ m/s, $\theta = 30°$ (a) $vw = 10$ m/s, $\theta = 120°$ (b)

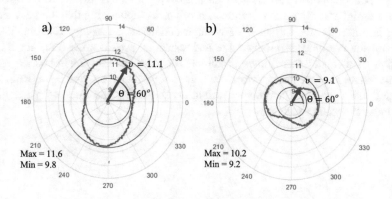

Fig. 3. Radar charts of CS resistance to changes in wind speed.

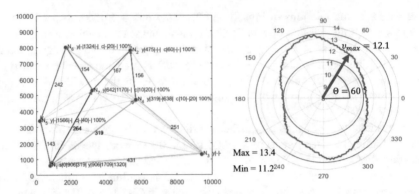

Fig. 4. Sub-mission with the highest resistance to changes in wind speed $vm = 11.2$ [m/s].

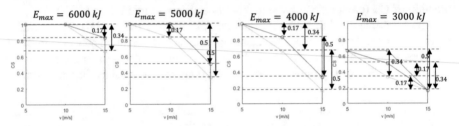

Fig. 5. Customer satisfaction as a function of wind speed

6   Conclusions

The declarative model proposed here (implemented in the ILGO IBM environment) allows one to determine UAV missions such that the customer satisfaction level is maximized under various weather conditions. The permissible size of the distribution network (12 nodes and 5 UAVs) for which such missions can be determined, makes the proposed model particularly suitable for application within an approach in which a network is decomposed into clusters, each covering a part of the set of all customers serviced during one flying time window. It is worth emphasizing that the possibility of taking into account the influence of weather conditions on energy consumption, and hence on the customer-servicing route and schedule, provides the basis for the construction of a model that allows searching for missions that are robust to specific weather changes. Our future work will, therefore focus on finding a method for choosing from among alternative distribution structures, the ones with the highest robustness to weather changes.

References

1. Adbelhafiz, M., Mostafa, A., Girard, A.: Vehicle routing problem instances: application to multi-UAV mission planning. In: AIAA Guidance, Navigation, and Control Conference. American Institute of Aeronautics and Astronautics (2010)
2. Belkadi, A., Abaunza, H., Ciarletta, L., Castillo, P., Theilliol, D.: Distributed path planning for controlling a fleet of UAVs: application to a team of quadrotors To cite this version: HAL Id: hal-01537777 Distributed Path Planning for Controlling a Fleet of UAVs: Application to a Team of. (2017)
3. Bocewicz, G., Nielsen, P., Banaszak, Z., Thibbotuwawa, A.: A declarative modelling framework for routing of multiple UAVs in a system with mobile battery swapping stations. In: Intelligent Systems in Production Engineering and Maintenance. ISPEM 2018. Advances in Intelligent Systems and Computing, vol. 835, pp. 429–441 (2018). https://doi.org/10.1007/978-3-319-97490-3_42
4. Bocewicz, G., Nielsen, P., Banaszak, Z., Thibbotuwawa, A.: Routing and scheduling of unmanned aerial vehicles subject to cyclic production flow constraints. In: Advances in Intelligent Systems and Computing, vol. 801, pp. 75–86 (2019). https://doi.org/10.1007/978-3-319-99608-0_9
5. Coelho, B.N., Coelho, V.N., Coelho, I.M., Ochi, L.S., Haghnazar, K.R., Zuidema, D., Lima, M.S.F., da Costa, A.R.: A multi-objective green UAV routing problem. Comput. Oper. Res. 0, 1–10 (2017). https://doi.org/10.1016/j.cor.2017.04.011
6. Dorling, K., Heinrichs, J., Messier, G.G., Magierowski, S.: Vehicle routing problems for drone delivery. IEEE Trans. Syst. Man Cybern. Syst. 47, 70–85 (2017). https://doi.org/10.1109/tsmc.2016.2582745
7. Drucker, N., Penn, M., Strichman, O.: Cyclic routing of unmanned aerial vehicles. In: Lect. Notes Comput. Sci. (including Subser. Lect. Notes Artif. Intell. Lect. Notes Bioinformatics), vol. 9676, pp. 125–141 (2016). https://doi.org/10.1007/978-3-319-33954-2_10
8. Geyer, C., Dey, D., Singh, S.: Prototype sense-and-avoid stemy for UAVs. Report (2009)
9. Gola, A., Kłosowski, G.: Application of fuzzy logic and genetic algorithms in automated works transport organization. In: Advances in Intelligent Systems and Computing, vol. 620, pp. 29–36 2018. https://doi.org/10.1007/978-3-319-62410-5_4
10. Goerzen, C., Kong, Z., Mettler, B.: A survey of motion planning algorithms from the perspective of autonomous UAV guidance (2010)
11. Habib, D., Jamal, H., Khan, S.A.: Employing multiple unmanned aerial vehicles for co-operative path planning. Int. J. Adv. Robot. Syst. 10, 1–9 (2013). https://doi.org/10.5772/56286
12. Khosiawan, Y., Nielsen, I., Do, N.A.D., Yahya, B.N.: Concept of indoor 3D-route UAV scheduling system. In: Advances in Intelligent Systems and Computing, pp. 79–88 (2016)
13. LaValle, S.M.: Planning Algorithms. Cambridge University Press, Cambridge (2006)
14. Tian, J., Shen, L., Zheng, Y.: Genetic algorithm based approach for multi-UAV cooperative reconnaissance mission planning problem. Presented at the BT - Foundations of Intelligent Systems (2006)
15. Liu, X.F., Guan, Z.W., Song, Y.Q., Chen, D.S.: An optimization model of UAV route planning for road segment surveillance. J. Cent. South Univ. 21, 2501–2510 (2014). https://doi.org/10.1007/s11771-014-2205-z
16. Rubio, J.C., Kragelund, S.: The trans-pacific crossing: long range adaptive path planning for UAVs through variable wind fields. In: The 22nd Digital Avionics Systems Conference, DASC 2003, p. 8–B. IEEE (2003)

17. Sitek, P., Wikarek, J.: Capacitated vehicle routing problem with pick-up and alternative delivery (CVRPPAD) – model and implementation using hybrid approach. Ann. Oper. Res. **273**, 257–277 (2019). https://doi.org/10.1007/s10479-017-2722-x
18. Thibbotuwawa, A., Nielsen, P., Banaszak Z., Bocewicz, G.: Energy consumption in unmanned aerial vehicles: a review of energy consumption models and their relation to the UAV routing. In: Advances in Intelligent Systems and Computing, vol. 853, pp. 173–184 (2019). https://doi.org/10.1007/978-3-319-99996-8_16

Graph of Primitives Matching Problem in the World Wide Web CBIR Searching Using Query by Approximate Shapes

Roman Stanisław Deniziak and Tomasz Michno[(✉)]

Kielce University of Technology,
al. Tysiaclecia Panstwa Polskiego 7, 25-314 Kielce, Poland
{s.deniziak,t.michno}@tu.kielce.pl

Abstract. The problem of matching two graphs has been considered by many researchers. Our research, the WWW CBIR searching using Query by Approximate Shapes is based on decomposing a query into a graph of primitives, which stores in nodes the type of a primitive with its attributes and in edges the mutual position relations of connected nodes. When the graphs of primitives are stored in a multimedia database used for World Wide Web CBIR searching, the methods of comparisons should be effective because of a very huge number of stored data. Finding such methods was a motivation for this research. In this initial research only the simplest methods are examined: NEH-based, random search-based and Greedy.

Keywords: WWW CBIR · Graphs matching · Graphs of primitives

1 Introduction

The problem of matching two graphs has been considered by many researchers. In the first half of the 21st century the first theorems by König were presented which were very important in the area of the graphs matching [8]. The recent researches trie to solve the problem of graph matching e.g. using incremental B-spline deformation [7], a belief propagation algorithm in order to minimize the energy function in the graph matching optimization problem [5] or Adaptive and Branching Path Following [9]. There are also methods based on ant colony optimization [10], Aggregated Search [1] or Deep Learning [2].

Our research, the World Wide Web CBIR searching using Query by Approximate Shapes presented in [3] is based on decomposing a query into a graph of primitives. The graph of primitives is an extended definition of a graph, where the nodes store the type of a primitive (a line segment, a polyline, a polygon, an arc, a chain of connected arches and a polygon built from arches) and its attributes. Moreover, edges store additional information like mutual position relations of connected nodes (using geographical directions) [4]. The difference between matching nodes between graphs and matching nodes between graphs

© Springer Nature Switzerland AG 2020
E. Herrera-Viedma et al. (Eds.): DCAI 2019, AISC 1004, pp. 77–84, 2020.
https://doi.org/10.1007/978-3-030-23946-6_9

of primitives includes checking the type of a primitive stored in a node and testing their attributes. Moreover, when edges are compared, the mutual primitives position relations should be also taken into consideration. For this reason, computing the similarity between graphs of primitives needs additional steps. Moreover, when the graphs of primitives are stored in a multimedia database used for World Wide Web CBIR searching, the methods of comparisons should be effective because of a very huge number of stored data [3]. Finding such methods was a motivation for this research. Since we found out that matching nodes in graphs of primitives is similar in some parts to solving the problem of optimal production planning and scheduling, there is a need of testing if such methods can be applied. In this initial research only the simplest methods are examined: NEH-based, Random search-based and Greedy.

The paper is organized as follows: the first section provides a brief introduction to the topic and the motivation of the research. In the second section, we present different algorithms for graphs of primitives nodes matching. The next section presents the experimental results. The fourth section provides the conclusion of the research and further development directions. The last section is a list of references.

2   The Problem of Matching Nodes Between Graphs of Primitives

The problem of proper nodes matching between graphs of primitives in the World Wide Web CBIR searching using Query by Approximate Shapes is very crucial. The similarity of two graphs of primitives is measured by the *similarity* coefficient which is defined as the maximum sum of each similarity between matched nodes divided by the minimum number of nodes in graphs [4]. The algorithm of measuring the similarity between nodes was described more detailed in [4].

The similarities between nodes of the compared two graphs may be stored in the *similarities matrix*, where nodes of the first graph are represented by rows and nodes of the second graph are represented by columns. Nodes of the graph with higher number of nodes is always represented by rows. The example matrix is presented in the Fig. 1 (the first graph contains 3 nodes, the second - 2 nodes). Thus, the problem of finding the *similarity* between graphs may be defined as the optimization problem, where for the matching of columns to rows their *similarities matrix* elements sum should be maximized.

Since, the problem of finding the maximum *similarity* value examining all possible matching between graphs nodes is very time consuming, there is a need of finding more efficient methods.

The problem is in some parts similar to production planning and scheduling problem, therefore the motivation of the research was to examine if such methods may be applied to nodes matching. In this initial research only the simplest methods are tested: brute-force, based on NEH algorithm, simulation methods and greedy algorithm approach. As the further development, more advanced methods will be examined.

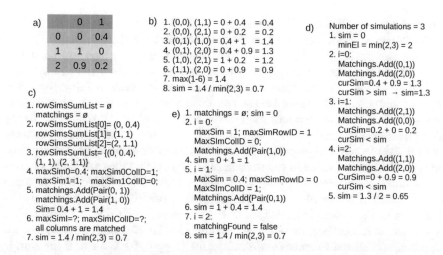

Fig. 1. The example maximum similarity matching algorithms executions: (a) the example *similarities matrix*, (b) Brute force algorithm, (c) NEH-based algorithm, (d) Random search-based algorithm, (e) Greedy algorithm.

2.1   Brute Force Algorithm

The Brute force algorithm tests all possible matching between graphs nodes and choose the highest *similarity* sum for them. Thus the obtained result is the most precise, but also the most time consuming. Due to the limited space, the algorithm pseudo-code is not presented. The example execution of the algorithm is shown in the Fig. 1(b). Firstly all permutations are found and for each of them the *similarities* matrix values are summed up (steps 1–6). Then, the maximum sum is chosen (step 7 - the maximum value is equal to 1.4) and in the step 8 it is divided by the minimum nodes number (2) giving the graphs similarity value equal to 0.7.

2.2   NEH-Based Algorithm

The NEH (Nawaz, Enscore, Ham) [6] algorithm is a deterministic algorithm which solves the problem of job scheduling sorting them descendant by their total processing time on machines. Next, two first jobs are chosen as the initial sequence and then others from the sorted list are inserted in order to minimize the total processing time of the whole sequence.

The problem of matching rows to columns in the graphs nodes *similarities matrix* may be solved using modified NEH algorithm. The modification includes changing the minimization problem into maximization, replacing the jobs and machines by graphs nodes (the nodes from graph with higher number of nodes is used as the replacement for jobs, the nodes with lower number by machines), changing the sorting order from descending into ascending and adding the constraint that each row and column may be used only once. The first algorithm step is to sum up all elements for each row and add them to the list. After that, the list is sorted ascending and two its first elements are chosen. Sorting highly improves the probability that nodes from the first graph which are harder to match to nodes of the second graph will be used at first. Then the column with the highest value of *similarities matrix* is chosen for each candidates row. If column numbers are the same for the row with lower value the another column is chosen. Then the found values from *matrix* are added to the graphs similarity value (*sim*) and the matched row-column pairs are stored. Next, other elements from the list are selected, choosing for them columns which have not been paired and with the highest value in the *matrix*. The algorithm stops when all columns are matched. After choosing matching, the *sim* value is divided by the minimum number of nodes in graphs. The algorithm pseudo code is shown in the Algorithm 1. The example execution of the algorithm is shown in the Fig. 1(c). Firstly, for each row the sum of corresponding *similarities matrix* elements are computed and stored in *rowSimsSumList* list (step 2, for the row 0 the sum is equal to 0.4, for the row 1 the sum is equal to 1 and for the row 2 the sum is equal 1.1). Next, all list elements are sorted (step 3). After that, the maximum values for two first list elements rows (number 0 and 1) are selected and stored as matching (steps 4 and 5). Also as the graphs *similarity sim* found maximum values are added (step 5, *sim* = 1.4). Next, in the step 6, there is tested if all columns are matched - in the example the test result is true. After that, in the step 7, the *sim* value is divided by the minimum number of graphs nodes (2), which results in graphs similarity equal to 0.7.

2.3   Random Search-Based Algorithm

The Random search algorithm finds only sub-optimal solution for the problem randomly selecting the matching for chosen number of trials. Firstly for each trial, all matching pairs are chosen using random number generator, considering the constraint of only once rows and columns usage. When the matching is selected, all *similarities matrix* values for them are summed up. If the computed value is the highest from each previous trials, it is stored. After all trials iterations, the found maximum value is divided by the minimum number of nodes in graphs. Due to the limited space, the algorithm pseudo-code is not presented. The example algorithm execution is shown in the Fig. 1(d). The total number of simulations is fixed to 3 due to the limited space, in practical implementations it should be much higher, e.g. 10 000. As the first simulation iteration (step 2), the node 0 from the first graph was matched to the node 1 and the node 2 to node 0 which resulted in its sum of similarity equal to 1.3 (which was stored as maximum). Next, the second iteration (step 3) matched node 2 to node 1 and node 0 to node 0, which resulted in sum equal to 0.2, which was lower than current maximum. The third iteration (step 4) matched node 1 to node 1 and node 2 to node 0, which resulted in sum equal to 0.9, which was also lower than current maximum. After all simulation steps, the maximum similarity *sim* is divided by the minimum number of nodes (2), which resulted in the graphs *similarity* equal to 0.65 (step 5).

Algorithm 1. NEH-based nodes matching algorithm.

Input: *similarities* - two-dimensional array storing the computed similarity between each nodes, G_i, G_j - compared graphs of primitives

```
 1: sim ← 0
 2: if nodes count of Gi or Gj ¡ 2 then
 3:    sim ← the matching with highest value using Brute Force
 4: else
 5:    rowsSimsSumList ← empty list of Pairs(int rowID, double sum)
 6:    matchings ← empty list of Pairs(int rowID, int columnID)
 7:    for i ← 0 to ((Gi nodes count) − 1) do
 8:       curSim ← 0
 9:       for j ← 0 to ((Gj nodes count) − 1) do
10:          curSim = curSim + similarities[matching.i][matching.j]
11:       end for
12:       rowsSimsSumList.Add(Pair(i, curSim))
13:    end for
14:    sort rowsSimsSumList list (ascendent)
15:    find the highest value in similarities for row rowsSimsSumList[0].rowID and store it in
       maxSim0, store also the column number in maxSim0ColID;
16:    find the highest value in similarities for row rowsSimsSumList[1].rowID and store it in
       maxSim1, store also the column number in maxSim1ColID;
17:    if maxSim0ColID <> maxSim1ColID then
18:       matchings.Add(Pair(rowsSimsSumList[0].rowID, maxSim0ColID))
19:       matchings.Add(Pair(rowsSimsSumList[1].rowID, maxSim1ColID))
20:       sim = maxSim0 + maxSim1
21:    else
22:       if maxSim0 > maxSim1 then
23:          sim = maxSim0
24:          matchings.Add(Pair(rowsSimsSumList[0].rowID, maxSim0ColID))
25:          find the highest value in similarities for row rowsSimsSumList[1].rowID (omitting
             column maxSim0ColID) and store it in maxSim1, store also the column number in
             maxSim1ColID;
26:          sim = sim + maxSim1
27:          matchings.Add(Pair(rowsSimsSumList[1].rowID, maxSim1ColID))
28:       else
29:          sim = maxSim1
30:          matchings.Add(Pair(rowsSimsSumList[1].rowID, maxSim1ColID))
31:          find the highest value in similarities for row rowsSimsSumList[0].rowID (omitting
             column maxSim1ColID) and store it in maxSim0, store also the column number in
             maxSim0ColID;
32:          sim = sim + maxSim0
33:          matchings.Add(Pair(rowsSimsSumList[0].rowID, maxSim0ColID))
34:       end if
35:    end if
36:    for i ← 2 to ((rowsSimsSumList elements count) − 1) do
37:       if all columns are matched then
38:          break;
39:       end if
40:       find the highest value in similarities for row rowsSimsSumList[i].rowID (omitting
          columns from matchings) and store it in maxSimI, store also the column number in
          maxSimIColID;
41:       sim = sim + maxSimI
42:       matchings.Add(Pair(rowsSimsSumList[i].rowID, maxSimIColID))
43:    end for
44: end if
45: sim ← sim/min(number of Gi nodes, number of Gj nodes)
46: if sim > 1 then
47:    sim ← 1
48: end if
49: return sim
```

2.4   Greedy Algorithm

The greedy algorithm tries to find the solution choosing the highest *similarity* values from the *similarities matrix* in order to choose the graphs nodes matching. That approach very often provides the same results as e.g. brute force algorithm, but it may also as the result find only sub-optimal solution. Due to the limited space, the algorithm pseudo-code is not presented. The example algorithm execution is shown in the Fig. 1(e). The algorithm is executed until all columns are not matched. Firstly, the maximum *similarity* value is chosen from the whole *similarities matrix* and row-column matching is added to *matchigs* (step 2). In the example, the maximum value is equal to 1 and is obtained for row number 1 and column number 0. Also, the found maximum value is added to the graphs nodes similarity *sim* value (step 4). Then, in another iteration, next maximum value from the *matrix* is chosen, omitting previously matched rows and columns and row-column matching is added to *matchigs* (step 5). In the example, the maximum value is equal to 0.4 and is obtained for row number 0 and column number 1. As previously, the maximum value is added to *sim*. In the next iteration, after checking if there are any not matched columns, the algorithm breaks the loop (step 7). After that, the maximum similarity *sim* is divided by the minimum number of nodes (2), which resulted in the graphs *similarity* equal to 0.7 (step 8).

3   The Experimental Results

The nodes matching algorithms were tested using an experimental application. The experiments were executed for small and moderate number of graph nodes in order to check the performance of the algorithms for each of them. Moreover, test for examining how different algorithms influence the precision and recall results of a query on the World Wide Web CBIR searching were performed.

First experiments were performed for graphs of primitives with small number of nodes (from 2 to 4 primitives). The brute force algorithm results may be used as a reference, because they check all possible matching. The computation times for algorithms were very similar and equal to 0 ms (except random search algorithm which executes in about 340 ms for 100 000 simulations). For the Brute force, Random search-based and Greedy algorithm all results were the same. Only the NEH-based algorithm for two graphs (no. 2 and 3) returned much lower values then others.

The second type of experiments were run for a moderate number of nodes (about 5). The results were very similar to the previous ones - the most precise values were obtained by the Brute-force, Random search-based and Greedy algorithms, but the Random search algorithm obtained a bit lower *similarity* for one graph the two others. The NEH-based algorithm obtained the lowest values. The execution times for NEH-based and Greedy algorithm were lower than millisecond, the Brute-force algorithm executed in about 56 min and the Random search-based between 554–607 ms. The test results for graphs are shown in the Table 1.

The third type of experiments were performed in order to test how different algorithms influence the precision and recall results of a query on the World Wide Web CBIR searching. For tests were used graphs of primitives extracted from photos with number of nodes from 25 to 150. Because of very long computation time the Brute-force method was not tested. The test results are shown in the Table 2. Contrary to the tests for small graphs of primitives, the NEH-based method gave the highest values of results precision and recall. Moreover, the computation time was also the shortest for that method.

Table 1. The experimental results of *similarity* between small (graphs 1–3) and medium (graphs 4–6) graphs.

Comparison with	Brute force			NEH-based			Random search-based			Greedy		
	gr. 1	gr. 2	gr. 3	gr. 1	gr. 2	gr. 3	gr. 1	gr. 2	gr. 3	gr. 1	gr. 2	gr. 3
Graph no. 1	1	1	0	1	1	0	1	1	0	1	1	0
Graph no. 2	1	1	0	1	0.67	0	1	1	0	1	1	0
Graph no. 3	0	0	1	0	0	0.5	0	0	1	0	0	1
Comparison with	Brute force			NEH-based			Random search-based			Greedy		
	gr. 4	gr. 5	gr. 6	gr. 4	gr. 5	gr. 6	gr. 4	gr. 5	gr. 6	gr. 4	gr. 5	gr. 6
Graph no. 4	1	0.51	0	0.4	0	0	1	0.51	0	1	0.51	0
Graph no. 5	0.64	1	0	0	0.48	0	0.51	1	0	0.64	1	0
Graph no. 6	0	0	1	0	0	0.37	0	0	1	0	0	1
The execution times for moderate graphs (ms)												
Comparison with	Brute force			NEH-based			Random search-based			Greedy		
	gr. 4	gr. 5	gr. 6	gr. 4	gr. 5	gr. 6	gr. 4	gr. 5	gr. 6	gr. 4	gr. 5	gr. 6
Graph no. 4	339396	339682	337433	0	0	0	603	566	554	0	0	0
Graph no. 5	336841	340331	335317	0	0	0	607	568	593	0	0	0
Graph no. 6	336551	340109	335270	0	0	0	599	562	554	0	0	0

Table 2. The experimental results of query by a bicycle and car graphs of primitives.

Comparison with	NEH-based			Random search-based			Greedy		
	Precision	Recall	Time (ms)	Precision	Recall	Time (ms)	Precision	Recall	Time (ms)
Bicycle	0.92	0.65	1.11	0.5	1	17598.13	0.38	0.47	169.24
Car	0.43	0.51	0.13	0.4	0.7	7107.06	0.41	0.33	21.67

4   Conclusions and Further Research Directions

This paper presented the initial research on solving the graph of primitives nodes matching for the World Wide Web CBIR searching using Query by Approximate Shapes. Four types of methods were presented and tested: Brute-force algorithm,

NEH-based algorithm, Random search-based algorithm and Greedy algorithm. The results are interesting especially for NEH-based algorithm, because for the query tests it provided the highest precision and recall results and the shortest computation time.

The further development of the research includes testing all presented algorithms on different classes of objects presented in images. Moreover more advanced methods should be examined which would provide more precise results. As an additional area of research, the database structure which stores the graphs of primitives should be improved.

References

1. A graph matching approach based on aggregated search. In: 2017 13th International Conference on 2017 13th International Conference on Signal-Image Technology & Internet-Based Systems (SITIS), Signal-Image Technology & Internet-Based Systems (SITIS), SITIS, p. 376 (2017)
2. Deep learning of graph matching. In: 2018 IEEE/CVF Conference on 2018 IEEE/CVF Conference on Computer Vision and Pattern Recognition, Computer Vision and Pattern Recognition (CVPR), CVPR, p. 2684 (2018)
3. Deniziak, R.S., Michno, T.: World wide web cbir searching using query by approximate shapes. In: Rodríguez, S., Prieto, J., Faria, P., Kłos, S., Fernández, A., Mazuelas, S., Jiménez-López, M.D., Moreno, M.N., Navarro, E.M. (eds.) Distributed Computing and Artificial Intelligence, Special Sessions, 15th International Conference, pp. 87–95. Springer, Cham (2019)
4. Deniziak, S., Michno, T.: New content based image retrieval database structure using query by approximate shapes. In: 2017 Federated Conference on Computer Science and Information Systems (FedCSIS), pp. 613–621, September 2017
5. Lin, X., Niu, D., Zhao, X., Yang, B., Zhang, C.: A novel method for graph matching based on belief propagation. Neurocomputing **325**, 131–141 (2019)
6. Nawaz, M., Enscore Jr., E.E., Ham, I.: A heuristic algorithm for the m-machine, n-job flow-shop sequencing problem. Omega **11**(1), 91–95 (1983)
7. Pinheiro, M.A., Kybic, J.: Incremental B-spline deformation model for geometric graph matching. In: 2018 IEEE 15th International Symposium on Biomedical Imaging (ISBI 2018), pp. 1079–1082, April 2018
8. Plummer, M.D.: Matching theory—a sampler: from Dénes König to the present. Discret. Math. **100**(1), 177–219 (1992)
9. Wang, T., Ling, H., Lang, C., Feng, S.: Graph matching with adaptive and branching path following. IEEE Trans. Pattern Anal. Mach. Intell. **40**(12), 2853–2867 (2018)
10. Wu, Y., Gong, M., Ma, W., Wang, S.: High-order graph matching based on ant colony optimization. Neurocomputing **328**, 97–104 (2019). Chinese Conference on Computer Vision 2017

Seabed Coverage Path Re-Routing
for an Autonomous Surface Vehicle

Andreas Hald Espensen[1], Oskar Emil Aver[1], Pernille Krog Poulsen[1],
Inkyung Sung[2(✉)], and Peter Nielsen[2]

[1] Department of Mathematics, Aalborg University, Aalborg, Denmark
{aespen14,oaver11,ppouls14}@student.aau.dk
[2] Operations Research Group, Department of Materials and Production,
Aalborg University, Aalborg, Denmark
{inkyung_sung,peter}@mp.aau.dk

Abstract. An Autonomous Surface Vehicle (ASV) equipped with a
sonar sensor is used to survey seabed. As a ASV sails an offshore region
guided by a coverage path, the unit collects sonar data utilized to map
the seabed of the region. However, a sonar sensor occasionally fails to
provide data with an adequate quality for the mapping. This calls the
necessity of re-routing compared to the original path assigned to an operating ASV, as the ASV needs to re-visit the areas where the sensor data
is inadequate. Motivated by the fact, we address the issues in the ASV
re-routing from scheduling perspective and propose an approach to solve
the problem.

Keywords: Autonomous Surface Vehicle (ASV) ·
Coverage path planning · Re-routing · Seabed mapping

1 A Coverage Path Re-Routing Problem in Seabed Mapping

Autonomous vehicles are increasingly used for a variety of tasks, including material transportation tasks, surveillance, inspection, etc., based on their effectiveness in cost and task executions [2,3,9,11]. With the successful implementations of the autonomous vehicles in many business sectors especially manufacturing, the application area of the units has been broadened including marine
surveys, e.g. seabed mapping. The demand for the marine surveys conducted by
Autonomous Surface Vehicles (ASVs) with sonar sensors is increasing as both
industry and armed forces realize the potentials of the unit.

One of the issues in marine surveys using ASVs is that the sonar used often
fails to provide survey data of an adequate quality. Consider a situation where
a ASV is moving following a path to cover a region of interest as described in
Fig. 1 (a). After surveying a part of the region, the sonar data obtained during

---

This study has been conducted in corporation with Teledyne RESON A/S.

E. Herrera-Viedma et al. (Eds.): DCAI 2019, AISC 1004, pp. 85–92, 2020.
https://doi.org/10.1007/978-3-030-23946-6_10

the survey is analysed and a control centre notifies the ASV with respect to the area where the sonar sensor data is inadequate to map the seabed (the shaded area in Fig. 1 (b)). In this case, the area with inadequate sensor data quality may need to be surveyed again by the ASV.

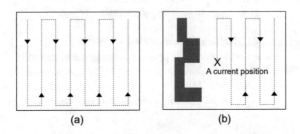

Fig. 1. Illustration for the ASV re-routing problem (a) an original coverage path (b) feedback about the poor quality of sensor data (shaded area)

A question arising at this situation is how to re-visit the area with low quality of sensor data while covering the area unexplored yet. The decision-making related with the situation is referred to as a ASV re-routing problem. While a coverage path planning problem seeking a path that guides a ASV to cover the area of interest in a cost and time effective way has been actively studied [4,5,8,10], the studies for the re-routing problem has been neglected.

The main differences between the coverage path planning and the ASV re-routing are as follows. First, the shape of area to survey differs between them. In general, the coverage path planning addresses a coverage path on a single polygon, whereas the re-routing often needs to find a coverage path on separating polygons as described in Fig. 1(b). Next, in the coverage path planning, start and end points of a coverage path are often same because a ASV needs to return to its original position (e.g. a home base). On the other hand, in the re-routing problem, a ASV does not necessarily need to return to its current position.

Importantly, these differences imply the fact that an algorithm designed for a coverage path planning is not directly applied to the re-routing problem thus an approach to solve the ASV re-routing problem should be designed considering the differences.

Motivated by the fact, this study addresses an approach to solve the re-routing problem based on the unique characteristics of the re-routing problem. Specifically, we first propose a way to find the connectivity between separating polygons by introducing a concept of *soft edge* where a ASV can cross. Given a connectivity network generated based on the concept, we then find a sequence of polygons with specific start and end polygons that allows a ASV to visit all polygons to survey with the minimum travel distance. Finally, a complete coverage path is formed based on the sequence. Note that the focus of this study is on a structure of the re-routing solving rather than on e.g. speeding up the problem solving for a large complex problem.

The remainder of this paper is structured as follows. Section 2 covers a general approach for a coverage path planning problem, which is a basis of the proposed approach for the re-routing problem. Section 3 describes the approach for the re-routing problem followed by concluding remarks of this study in Section 4.

2   Coverage Path Planning

2.1   Convex Polygon

Consider a convex polygon as an area to be covered using a ASV. In this case, a coverage path can be easily found as parallel tracks/lanes to one of the edges of the polygon. The boustrophedon method [4] is rooted in this idea. Figure 2 illustrates the concept of the method.

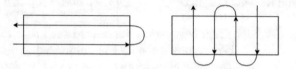

Fig. 2. Boustrophedon method

Note that a common criteria to measure the performance of a coverage path found by applying the boustrophedon method is the number of turns in the path. This is due to that turns lead to a longer travel distance and the quality of sensor data gathered during the turn is often poor. Thus, the coverage path on the left in Fig. 2 is preferred than the path on the right.

To find a coverage path with the minimum numbers of turns, let us first introduce definitions about a span and a width of a convex polygon following the work of [8] as follows:

Definition 1. *The span of a convex polygon is the distance between a pair of support parallel lines.*

Definition 2. *The width of a convex polygon is the minimum span.*

With the definitions, the number of turns of a coverage path for a convex polygon found by the boustrophedon method is minimized when the direction of the path is perpendicular to the direction of the width of the polygon. Please refer to [6,12] for the details of the concept and algorithms that find a width of a convex polygon.

2.2   Concave Polygon

In case of concave polygons, the way of generating a coverage path with a single direction is not suitable. As illustrated in Fig. 3, a shorter coverage path is obtained by assigning different directions to the path.

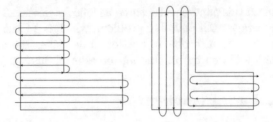

Fig. 3. A drawback of having a single direction for a coverage path on a concave polygon

Following the fact, the first step to find a coverage path over a concave polygon is to decompose the polygon into convex sub-polygons. A Morse-based cellular [1,5] and a sweep line algorithm [7,12] have been proposed for the convex decomposition.

The convex decomposition is generally conducted such that the total sum of widths of the convex sub-polygons generated is minimized. This can minimize the total length of a final coverage plan. More formally, let P denote a non-convex polygon with n vertices. A convex decomposition of P consists of $m \leq n$ convex sub-polygons $P_1, P_2, \ldots, P_m$ with corresponding widths $W_1, W_2, \ldots, W_m$. Then the coverage problem can be stated as:

$$\min_{W_1, W_2, \ldots, W_m} \sum_{i=1}^{m} W_i \qquad (1)$$

$$\text{s.t } P_i \in \text{Convex Polygon.}$$

A convex polygon is a special case of (1), where $m = 1$.

The boustrophedon method is then applied to each convex sub-polygon to find a coverage path for the sub-polygon. Finally, a complete coverage path for the original concave polygon is generated by connecting the coverage paths for the sub-polygons with respect to minimizing the total distance of the path considering the connectivity between the sub-polygons. The procedures of obtaining a coverage path on a concave polygon is described in Fig. 4.

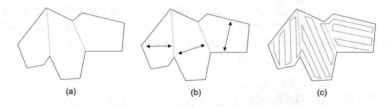

Fig. 4. The procedures of a coverage path planning (a) Convex decomposition (b) Finding widths for sub-polygons (c) Connecting sub-coverage paths

Lastly, it should be noted that the problem, which finds a sequence of sub-polygons to connect the sub-coverage paths as a complete coverage path, is generally formulated as a Traveling Salesman Problem (TSP) or finding a Hamiltonian walk because a ASV needs to cover all areas and return to its original position [12]. A walk is a sequence of nodes in a graph. A repetition of a node is allowed to form a walk.

3   ASV Re-Routing

3.1   Connecting Separating Polygons

Recall that one of the factors making the re-routing problem for a ASV distinguishable from the coverage path planning is a disjoint between areas to survey. In the re-routing, the areas (i.e. polygons) to survey are not necessarily connected. Furthermore, as the time to detect the lack of sensor data quality is delayed, the distance between the separating areas becomes large. Importantly, as explained in Sect. 2, the approach for the coverage path planning addresses a coverage path on a single polygon; thus, the approach cannot be directly applied to the re-routing problem.

To tackle the issue, we introduce a *soft edge* to connect separated polygons as illustrated in Fig. 5. In Fig. 5, two polygons are linked by soft edges represented as dotted lines. A boundary of a polygon of interest is considered as a soft edge. Based on the concept, two separating polygons are considered connected when a straight line between the polygons only crosses soft edges. A hard edge is also introduced to represent obstacles or a region where a ASV cannot sail for other reasons. When any straight line between two separating polygons always crosses a hard edge, the polygons are considered disconnected.

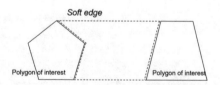

Fig. 5. Concept of the soft edge

3.2   Polygon Sequencing

An important feature of the ASV re-routing is that the unit does not necessarily need to return to its current position. Instead, the unit may need to be at a specific position at the end of survey e.g. its home base. Therefore, the polygon sequencing problem in the re-routing can be formulated as a problem that given specific start and end polygons of the sequence finds a sequence of polygons, which allows a ASV to visit all polygons with the minimum travel distance.

Let us consider graph G, whose nodes in set V and arcs in set E correspond to sub-polygons resulted from the convex decomposition and connectivity between the sub-polygons, respectively. Then a solution to the polygon sequencing problem is formally described as a $s - t$ walk traversing a graph (i.e. visiting all nodes), where s and t are start and end nodes of the walk, respectively.

To solve the problem, we propose a procedure that finds a walk from node s to node t traversing a graph with the minimum number of arcs. In the procedure, given a small number for k (e.g. the number of nodes in a graph), an algorithm, which seeks all $s - t$ walks with exactly k arcs, is executed. The algorithm is described in Algorithm 1. If the algorithm finds $s - t$ walks with k arcs traversing graph G, the procedure is terminated reporting the $s - t$ walk with the shortest distance. Otherwise, the algorithm is executed with the increased k value by one and the presence of a walk traversing graph G is checked again. These steps are repeated until a $s - t$ walk traversing graph G is found.

Algorithm 1. Find all walks from s to t with k arcs in $G = (V, E)$.

List $W_{i,j,e} = \varnothing$ $\forall i, j \in V, e \in 0, 1, ..., k$
for $e = 1, e \le k$ **do**
  for $i \in V$ **do**
    for $j \in V$ **do**
      if $e = 1$ and $(i, j) \in E$ **then**
        add (i, j) to $W_{i,j,1}$
      end if
      if $e > 1$ **then**
        for $a \in N$ **do**
          if $(i, a) \in E$ **then**
            $W_{i,j,e} = W_{a,j,e-1} \bigoplus (i, a)$
          end if
        end for
      end if
    end for
  end for
end for
Return $W_{s,t,k}$

In Algorithm 1, the function $W \bigoplus (i, j)$ is a function that combines all walks in set W with a walk (i, j). Specifically, the function inserts (i, j) at the fronts of all walks in W and returns the updated walk set W'. For example, $\{(a - b - c), (a - d)\} \bigoplus (c, a) = \{(c - a - b - c), (c - a - d)\}$.

3.3   An Approach for the ASV Re-Routing

Building upon the re-defined concept of the connectivity between polygons and the procedure to sequence polygons, an approach for the re-routing problem can be structured as follows.

Step 1. Given a set of separating polygons to survey, the convex decomposition is first applied to each individual polygon.

Step 2. The convex sub-polygons found from the convex decomposition is then used to determine the connectivity between them based on the concept of the soft and hard edges.

Step 3. $s - t$ walk with the minimum arcs that corresponds to a sequence of sub-polygons to visit is derived by the procedure combined with Algorithm 1.

Step 4. A complete coverage path is finally formed by connecting the sub-coverage paths for each sub-polygon generated by the boustrophedon method following the sequence of sub-polygons.

Figure 6 illustrates an example of a coverage path generated based on the approach. In the figure, a path represented as a line covers three separating polygons while avoiding a sail over the shaded area, which is an obstacle.

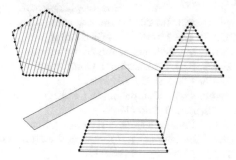

Fig. 6. A coverage path for three separating polygons

4   Concluding Remarks

In this study, we propose a solution structure for the ASV re-routing and its component algorithms based on the challenges in the problem solving. These are designed based on the fundamental differences between the coverage path planning and the ASV re-routing from the scheduling perspective. The proposed structure and algorithms can be a basis of a decision support system for a coverage path planning and re-routing. The performance of the proposed solution structure and algorithms will be further investigated especially in terms of the ability to generate a coverage path on-line.

It should be noted that the width of sonar beams of a ASV, which is assumed fixed in this study, can be further considered in the ASV re-routing. The width of sonar beams changes depending on the depth of sea and this is an important factor in determining a coverage path of a ASV. As demonstrated in [5], a sailing direction of a ASV can be determined as the direction perpendicular to the underlying seabed gradient so that the width of sonar beams can be relatively maintained at a consistent level over a survey mission.

Acquiring such information in advance of deriving an original coverage path can be difficult especially when a survey on a target area is scheduled for the first time. In case of the re-routing, however, there are available data obtained during a current survey mission, which can be used for the ASV re-routing.

References

1. Acar, E.U., Choset, H., Rizzi, A.A., Atkar, P.N., Hull, D.: Morse decompositions for coverage tasks. Int. J. Robot. Res. **21**(4), 331–344 (2002)
2. Bocewicz, G., Nielsen, I., Banaszak, Z.: Automated guided vehicles fleet match-up scheduling with production flow constraints. Eng. Appl. Artif. Intell. **30**, 49–62 (2014)
3. Bocewicz, G., Nielsen, P., Banaszak, Z., Thibbotuwawa, A.: Deployment of battery swapping stations for unmanned aerial vehicles subject to cyclic production flow constraints. Commun. Comput. Inf. Sci. **920**, 73–87 (2018)
4. Coombes, M., Chen, W.H., Liu, C.: Boustrophedon coverage path planning for UAV aerial surveys in wind. In: 2017 International Conference on Unmanned Aircraft Systems (ICUAS), pp. 1563–1571. IEEE (2017)
5. Galceran, E., Carreras, M.: Efficient seabed coverage path planning for ASVs and AUVs. In: 2012 IEEE/RSJ International Conference on Intelligent Robots and Systems (IROS), pp. 88–93. IEEE (2012)
6. Houle, M.E., Toussaint, G.T.: Computing the width of a set. IEEE Trans. Pattern Anal. Mach. Intell. **10**(5), 761–765 (1988)
7. Huang, W.H.: Optimal line-sweep-based decompositions for coverage algorithms. In: IEEE International Conference on Robotics and Automation, 2001. Proceedings 2001 ICRA, vol. 1, pp. 27–32. IEEE (2001)
8. Jiao, Y.S., Wang, X.M., Chen, H., Li, Y.: Research on the coverage path planning of UAVs for polygon areas. In: 2010 the 5th IEEE Conference on Industrial Electronics and Applications (ICIEA), pp. 1467–1472. IEEE (2010)
9. Khosiawan, Y., Khalfay, A., Nielsen, I.: Scheduling unmanned aerial vehicle and automated guided vehicle operations in an indoor manufacturing environment using differential evolution-fused particle swarm optimization. Int. J. Adv. Robot. Syst. **15**(1), 1–15 (2018)
10. Larson, J., Bruch, M., Ebken, J.: Autonomous navigation and obstacle avoidance for unmanned surface vehicles. In: Unmanned Systems Technology VIII, vol. 6230, p. 623007. International Society for Optics and Photonics (2006)
11. Liu, P., et al.: A review of rotorcraft unmanned aerial vehicle (UAV) developments and applications in civil engineering. Smart Struct. Syst **13**(6), 1065–1094 (2014)
12. Yu, X.: Optimization approaches for a dubins vehicle in coverage planning problem and traveling salesman problems. Ph.D. thesis (2015)

Multithreaded Application Model

Damian Giebas and Rafał Wojszczyk$^{(\boxtimes)}$

Faculty of Electronics and Computer Science, Koszalin University of Technology,
Koszalin, Poland
damian.giebas@s.tu.koszalin.pl, rafal.wojszczyk@tu.koszalin.pl

Abstract. The work presents an original model created for the detection of undesirable phenomena occurring in multithreaded applications. This model is characterized by taking time as one of the variables and dividing operations and resources into two separate entities. Based on the created model, the claim regarding the race condition was presented, and a proof confirming this claim was made.

Keywords: Multithreaded application model · Correctness criterion · Race condition

1 Introduction

Classic logic (i.e. predicate calculus) is the field of mathematics on which all modern programming languages are built and developers use them to create multithreaded applications. The correct multithreaded application, because applications are not always correct, should be understood as an application free from *race condition, deadlocks, atomicity violation* and *order violation*. This phenomena are commonly found in multithreaded applications and are often not noticed for many years because the situation in which they appear only in specific conditions. Finding snippets of the source code that cause these phenomena requires a lot of work during the code review. However, it can be much less time-consuming when the source code is brought e.g. to graphical representations. Paper [1] presents graphical representations of multithreaded applications. The Control Flow Graf (CFG) model distinguished the code bugs in the application, and the models in the form of Petri Net (PN) and Concurrent Process Systems (CPS) distinguished in the order of operations performed. None of these solutions treated operations and resources as separate entities, and also did not include time intervals as one of the variables. The division into operations and resources allows to unambiguously determine which operations performed in parallel use shared resources. Thanks to this knowledge it is possible to detect race condition. However, considering the time intervals allows to determine whether the threads of the application work in parallel, and what is related to it, if a number of conditions guaranteeing the correctness of the application are not met, the phenomena described in the paragraph above may occur.

E. Herrera-Viedma et al. (Eds.): DCAI 2019, AISC 1004, pp. 93–103, 2020.
https://doi.org/10.1007/978-3-030-23946-6_11

The author's graphic notation used in this article includes mutexes, which are used to ensure correctness in multithreaded applications. Thanks to mutexes, it is possible to counteract, for example, race condition, because operations using a shared resource, covered in the source code with the same block, preclude parallel use of the resource. Taking into account the mutexes in the graphical representation allows you to specify the minimum number of mutexes, which is necessary for the application to work properly. Determining the minimum number of mutexes makes it possible to reduce unnecessary mutexes, and what is associated with it, the chance of a deadlock is reduced.

This article presents the author's multithreaded application model that identifies errors in the source code structure of the program, which cause race condition. The Sect. 2 presents selected related works and the Sect. 3 contains a description of the proposed model and an illustrative example. Then, the Sects. 3 and 4 form the problem and describe the condition sufficiently. The Sect. 6 contains a summary of the work.

2   State of Art

In paper [6] authors present two reasons that the developers are still unable to cope with the phenomena described above. An important reason for this is the problem with collecting data concerning these phenomena. The apparent randomness of the multithreaded applications' operation makes it difficult for users to report bugs, and programmers to understand them. The apparent randomness is understood as the way the application works, in which one of the variables is time, and the application can take one of many states affected by the current state of the environment in which the application is being executed. In addition, multithreaded application errors are not easy to understand due to their structure of these phenomena. Their symptoms usually involve complicated interactions between many components of the program.

One of the methods of checking the correctness of multithreaded applications is the used of temporal logic. A description of this method can be found in, among others in paper [3]. This work was used to create training materials by Greniewski [4]. These materials show how to use the PROMELA language to build a model of the distributed system and verify it. However, this solution has significant disadvantages because, apart from writing the application, the programmer must also write the PROMELA code. It is also worth paying attention to the increase of the cost of maintaining the application, because along with its development, the code written in the PROMELA should also be maintaining.

Another way to validate mentioned applications is to use tools to analyze application binary files. A set of such tools has been discussed in paper [1]. The discussed tools for checking binary applications usually support a limited number of architectures, while the use of libraries and extensions increases the complexity of the code, which makes its analysis difficult, and very often it should also be reckoned with the possibility of abandoning their support.

There is also an approach based on the use of tools interfering with libraries and extensions to programming multithreaded applications. These tools change the implementation of selected functions of the libraries and extensions used, and do not change the application code. An example of such a tool is the *Grace* program discussed in paper [6]. Changes made by this tool result in the child processes being replaced by threads. Programs that use this tool are in most cases slower [5] than their counterparts using the *pthread* library, which is the result of Grace's method of operation. However, in return, this tool guarantees that no undesirable phenomenon will occur in the application. Among the drawbacks of this solution it is worth mentioning support of only 32 bit architecture, limitation of operation only to systems from the Linux family, significant slowdown of the application when resources shared by multiple threads change and no support for threads whose aim is to endlessly act, e.g. concurrent server.

In paper [7] authors describe the use of the PN to detect the race condition, deadlock, and also the lost signals bugs in event-driven programming. The authors did not take into account the disadvantages described in paper [1], i.e. at the moment when at least two transitions are started at the same time, there is no situation in which the wrong number of tokens appears at the target place. They also do not present how to read such a phenomenon from the models they generate. In addition, the authors of the work [7] presented mutexes in a different way than they did in the work [1], i.e. according to their claim for two threads, the mutex is implemented through one place with one token. Such a representation does not impose the order of work of given threads and does not cause mutual exclusion, as it was the case in the work [1]. This results in the fact that parallel working identical threads always have the same state at the selected moment of time. This leads to the rejection of a group of application states in which parallel threads have different states, and as a result, it is possible to reject the application state in which race condition or deadlock occurs.

Paper [6] is a comprehensive study of errors related to concurrency in computer applications. It focused on 105 known errors in multithreaded applications from programs: MySQL, Apache, Mozilla, and OpenOffice. The aim of the work was to examine several features of the application, including the structures and symptoms of errors, as well as strategies for their elimination. The further work of [6] indicate way on new tools for detecting undesirable phenomena in multithreaded applications, with the emphasis on multiple-variable bugs, as well as errors resulting from the order violation.

The analysis presented above shows that there is still no universal method to detect phenomena such as race condition, deadlock, atomicity violation and order violation.

3   Model

One of the representations used to model multithreaded applications is the PN. As shown in paper [1], multithreaded application models based on PN do not explicitly refer to source code elements of the application and do not take into account the changing dependencies between threads. This makes it difficult to develop a unified representation that allows detecting undesirable phenomena, regardless of the time in which they may occur.

Taking into account the aforementioned drawbacks, a model was proposed in which the multithreaded application P can be expressed in the form of the following n:

$$C_P = (T_P, U_P, R_P, O_P, \mathbb{M}_P, F_P),   \tag{1}$$

where:

1. $T_P = \{t_i | i = 0...\alpha\}$, $(\alpha \in \mathbb{N})$ is the set of threads t_i application P, where t_0 is the main thread, $|T_P| > 1$,
2. $U_P = (u_b | b = 1...\beta)$, $(\beta \in \mathbb{N}^+)$ is a sequence of sets u_b, which are subsets of a set of T_P that contain threads working in the same time interval in application P, with $|U_P| > 2$, $u_1 = \{t_0\}$ i $u_\beta = \{t_0\}$,
3. $R_P = \{r_c | c = 1...\gamma\}$, $(\gamma \in \mathbb{N}^+)$, $r_c = (v_c, w_c)$ is a set of shared resources in application P, defined as a pair *(variable, value)*,
4. $O_P = \{o_{i,j}, | i = 1...\delta, j = 1...\epsilon\}$, $(\delta, \epsilon, \in \mathbb{N}^+)$ is a set of all operations in application P, which at a certain level of abstraction are atomic operations, i.e. divide them into smaller operations. The operation should be understood as an instruction or function defined in the programming language. The index i indicates the number of the thread in which the operation occurs, and the index j is the ordinal number of operations working within the same thread.
5. $\mathbb{M}_P = (MU^b | b = 1...\beta)$ - sequence of MU_P^b; $MU_P^b = \{m_l^b | l = 1...\theta^b\}$ set of m_l^b mutexes valid for threads from the u_b; $m_l^b = \{o_{i,j} | t_i \in u_b\}$ - set of opposed operations (in the further part of this work the set will be called "mutex"). In addition, the following functions are defined for each m_l^b block:
   (a) $\phi_l^b(o_{i,j}) = \phi_{l,i,j}^b \in \mathbb{N}$ - the value of which is determined by the order in which l's was founded mutex for the operation $o_{i,j}$,
   (b) $\psi_l^b(o_{i,j}) = \psi_{l,i,j}^b \in \mathbb{N}$ - the value of which is determined by the order of unlocking l-th mutex for the operation $o_{i,j}$.
6. $F_P = \{f_n | n = 1...\iota\}$ i $F \subseteq (O_P \times O_P) \cup (O_P \times R_P)$, $(\iota \in \mathbb{N}^+)$ - set of edges including:
   (a) transition edges - defining the order of operations. These edges are pairs $f_n = (o_{i,j}, o_{i,k})$, where the elements describe two consecutive operations $o_{i,j} \in O_P$,
   (b) usage edges - indicating resources that change during the operation. These edges are pairs $f_n = (o_{i,j}, r_c)$, in which the first element is the operation $o_{i,j} \in O_P$, and the second element is the resource $r_c \in R_P$,
   (c) dependency edges - indicating operations dependent on the current value of one of the resources, so that the operation is performed correctly. These edges are pairs i $f_n = (r_c, o_{i,j})$, in which the first element is the resource $r_c \in R_P$, and the second element is the operation $o_{i,j} \in O_P$.

Figure 1 shows the sample application model (https://goo.gl/hiU1zH - written in C). The presented application contains 7 threads: t0, t1, ..., t6, which work in 7 time intervals: u1, u2, ..., u7. In each of the predefined time periods, *pthread_create* and *pthread_join* functions are called (the exception here is the interval u_5, which is the result of calling *pthread_cond_wait*). Threads of the application work on shared resources (6 resources have been identified), performing

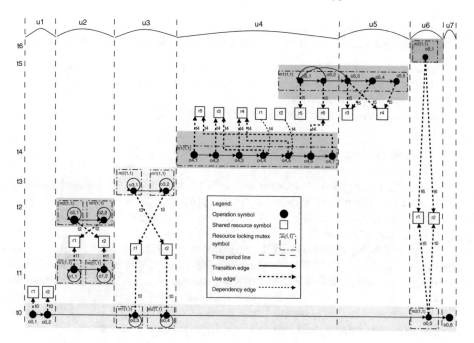

Fig. 1. Graphical representation of AP1 using the proposed notation.

operations: allocating memory on the heap for resources, modifying resources, and freeing memory allocated to resources.

According to the $C_P 1$ notation, the application under consideration (C_{AP1}) can be represented as follows:

$$T_{AP1} = \{t_0, t_1, t_2, t_3, t_4, t_5, t_6\},$$

$$U_{AP1} = (\{t_0\}, \{t_1, t_2\}, \{t_0, t_3\}, \{t_4, t_5\}, \{t_5\}, \{t_6, t_0\}, \{t_0\}),$$

$$R_{AP1} = \{(r1, 0), (r2, 0), (r3, 50), (r4, 70), (r5, w1), (r6, w2)\},$$

$$O_{AP1} = \{o_{0,4}, o_{0,2}, o_{0,3}, o_{0,4}, o_{0,5}, o_{0,6}, o_{1,1}, o_{1,2}, o_{2,1}, o_{2,2}, o_{3,1}, o_{3,2}, o_{4,1}, o_{4,2},$$
$$o_{4,3}, o_{4,4}, o_{4,5}, o_{4,6}, o_{4,7}, o_{5,1}, o_{5,2}, o_{5,3}, o_{5,4}, o_{5,5}, o_{6,1}\},$$

$$MU^2_{AP1} = \{\{o_{1,1}, o_{2,2}\}, \{o_{1,2}, o_{2,1}\}\}$$

$$MU^3_{AP1} = \{\{o_{0,3}, o_{3,2}\}, \{o_{0,4}, o_{3,1}\}\}$$

$$MU^4_{AP1} = \{\{o_{4,1}, o_{4,2}, o_{4,3}, o_{4,4}, o_{4,5}, o_{4,6}, o_{4,7}, o_{5,1}, o_{5,2}\}\}$$

$$MU^5_{AP1} = \{\{o_{5,3}, o_{5,4}, o_{5,5}\}\}$$

$$MU^6_{AP1} = \{\{o_{0,5}, o_{6,1}\}\}$$

$$F_{AP1} = \{(o_{0,1}, r_1), (o_{0,1}, o_{0,2}), (o_{o,2}, r_2), (o_{1,1}, r_1), (o_{1,1}, o_{1,2}), (o_{1,2}, r_2), (o_{2,1}, r_2),$$
$$(o_{2,1}, o_{2,2}), (o_{2,2}, r_1), (o_{0,2}, o_{0,3}), (o_{0,3}, r_1), (o_{0,3}, o_{0,4}), (o_{0,4}, r_2), (o_{3,1}, r_2),$$
$$(o_{3,1}, o_{3,2}), (o_{3,2}, r_1), (o_{4,1}, r_5), (o_{4,1}, o_{4,2}), (o_{4,2}, r_3), (o_{4,2}, o_{4,3}), (o_{4,3}, r_4),$$
$$(o_{4,3}, o_{4,4}), (r_1, o_{4,4}), (o_{4,4}, r_3), (o_{4,4}, o_{4,5}), (r_2, o_{4,5}), (o_{4,5}, r_4), (o_{4,5}, o_{4,6}),$$
$$(o_{4,6}, r_6), (o_{4,6}, o_{4,7}), (o_{4,7}, r_5), (o_{5,1}, r_5), (o_{5,1}, o_{5,2}), (o_{5,2}, r_6), (o_{5,2}, o_{5,3}),$$
$$(o_{5,3}, r_3), (o_{5,3}, r_4), (o_{5,3}, o_{5,4}), (o_{5,4}, r_3), (o_{5,4}, o_{5,5}), (o_{5,5}, r_4), (o_{0,4}, o_{0,5}),$$
$$(o_{0,5}, r_1), (o_{0,5}, r_2), (o_{6,1}, r_1), (o_{6,1}, r_2), (o_{0,5}, o_{0,6})\}.$$

Some of the elements present in the above model are not linked directly to the instructions present in the code and all these elements were described below:

1. the first element of the resource r_5, r5 means the *should_wait* variable,
2. the first element of the resource r_6, r6 means the *cond_var* variable,
3. w1 means *true* Boolean value,
4. w2 is the value given by the macro *PTHREAD_COND_INITIALIZER*.

4   Formulating the Problem

There is a multithreaded P application (written in C using the *pthread* library) with a race condition bug. Application A is represented in model form C_A 1. The answer to the question is: Is it possible to detect a race condition in application A based on the C_A model?

As an example, let us use RC1 applications (https://goo.gl/rEbdGw) and RC2 (https://goo.gl/y13TT9). In both applications there is a race condition resulting from uncontrolled access to the resource of operations performed in parallel working threads. According to the above question, a response is sought: Can a race condition be detected based on the proposed C_A 1 notation? In the next section, a condition will be described to enable such assessment.

5   Sufficient Condition

The race condition occurs when several processes concurrently reach for the same data and perform actions on them, as a result of which the result of these activities depends on the order in which the data was accessed [2]. In other words, race condition occurs when, at the same time, at least one of the operations in parallel working threads has uncontrolled access to the shared resource, is meant the lack of mutex. The RC1 application is the simplest example of a application with race condition. The model for this application C_{RC1} looks as follows:

$$T_{RC1} = (t_0, t_1, t_2), \qquad\qquad \mathbb{M} = \emptyset,$$
$$U_{RC1} = (\{t_0\}, \{t_1, t_2\}, \{t_0\}), \qquad F_{RC1} = \{(o_{0,1}, r_1), (o_{1,1}, o_{1,1}),$$
$$R_{RC1} = \{(r1, 0)\}, \qquad\qquad (o_{1,1}, r_1), (o_{2,1}, o_{2,1}), (o_{2,1}, r_1),$$
$$O_{RC1} = \{o_{0,1}, o_{0,2}, o_{1,1}, o_{2,1}\}, \qquad (o_{0,1}, o_{0,2}), (o_{0,2}, r_1)\}.$$

A graphic representation of the above model is shown in Fig. 2a. The application contains two threads t_1 and t_2 that have uncontrolled access to the resource r_1 (the set of $\mathbb{M}_{RC1}$ mutexes is empty). The consequence of the lack of control over access to shared resources is precisely the race condition, which is caused in particular by:

1. existence of two threads $u_2 = \{t_1, t_2\}$ working in the same time interval $u_2 \in U_{RC1}$,
2. existence of operations $o_{1,1}, o_{2,1} \in O_{RC1}$, which in parallel change the value of resource r_1,

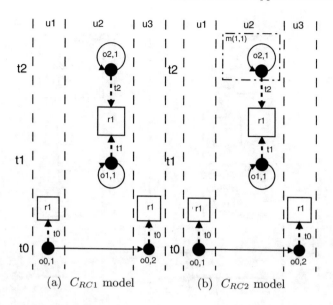

(a) C_{RC1} model   (b) C_{RC2} model

Fig. 2. Graphical representation of application models with the race condition.

3. and no mutex $m_l^b \in \mathbb{M}_{RC1}$, which excludes parallel work on the resource r_1.

In the RC2 application, as in RC1, there is also the race condition. Unlike RC1, however, the operation on a shared resource of one of the RC2 application threads is locked by a mutex. The race condition occurs due to uncontrolled access to the shared resource by the thread t_1. The C_{RC2} model looks as follows:

$$T_{RC2} = (t_0, t_1, t_2),$$
$$U_{RC2} = (\{t_0\}, \{t_1, t_2\}, \{t_0\}),$$
$$R_{RC2} = \{(r1, 0)\},$$
$$O_{RC2} = \{o_{0,1}, o_{0,2}, o_{1,1}, o_{2,1}\},$$

$$MU_{RC2}^2 = \{\{o_{2,1}\}\},$$
$$F_{RC2} = \{(o_{0,1}, r_1), (o_{1,1}, o_{1,1}),$$
$$(o_{1,1}, r_1), (o_{2,1}, o_{2,1}), (o_{2,1}, r_1),$$
$$(o_{0,1}, o_{0,2}), (o_{0,2}, r_1)\}.$$

The graphical representation of the RC2 application is shown in Fig. 2b. Unlike RC1, this model has the $m_1^2 = \{o_{2,1}\}$ mutex. The mutex $m_1^2 \in MU_{RC2}^2$ does not contain all operations $o_{i,j}$, using the resource $r_1 \in R_{RC2}$ (i.e. there exists an edge $f_n \in F_{RC2}$ connecting one of the operations $o_{i,j} \in O_{RC2}$ with the resource $r_1 \in R_{RC2}$),), and thus uncontrolled access is still present. In the case of RC2 application, the race condition occurs because:

1. there is a pair of $u_2 = \{t_1, t_2\}$ threads working in the same time interval $u_2 \in U_{RC2}$,
2. they carry out operations $o_{1,1}, o_{2,1} \in O_{RC2}$ on the shared resource r_1,
3. there is no $m_l^b \in MU_{RC2}^2$ mutex containing a set of operations $\{o_{1,1}, o_{2,1}\}$ working on the resource r_1.

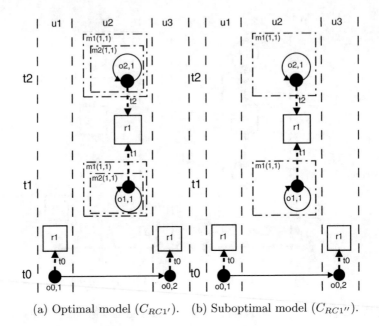

(a) Optimal model ($C_{RC1'}$).   (b) Suboptimal model ($C_{RC1''}$).

Fig. 3. Graphical representation of the C_{RC1} model without race condition.

Figure 3 contains two graphical representations with the solution to the problem of race condition of the C_{RC1} model. The solution ($C_{RC1'}$) from Fig. 3a is suboptimal, i.e. there is an excess mutex in the model $m_2 \in MU^2_{RC1}$, the removal of which results in a graphical representation as in Fig. 3b. Therefore, the second solution ($C_{RC1''}$), whose graphical representation is shown in Fig. 3b is optimal because it contains a sufficient number of mutexes that guarantees the correct operation of the multithreaded application.

The aforementioned reasons for the occurrence of race condition in multithreaded applications can be generalized to the following theorem:

Theorem 1. *Let $O_P = \{o_{m,j}, ..., o_{n,q}\}$ mean the set of operations performed within the threads of the $u_b \in U_P$ set, using the common resources $R_P = \{r_c, ..., r_k\}$ (i.e. $(o_{m,j}, r_c) \in F_P$ or $(r_c, o_{m,j}) \in F_P$, ..., or $(o_{n,q}, r_k) \in F_P$ or $(r_k, o_{n,q}) \in F_P$).*

If there exists such a two-element $O_P^* \subseteq O_P$ set, which is not a subset of any $m_l^b \in MU_P^b$ mutex (i.e. $\forall m_l^b \in MU_P^b : O_P^* \not\subseteq m_l^b$) there is a race condition between the O_P set operation.

Race condition identification is not the only application of the model. Another feature of the model is the distinction of information about the locking and unlocking of mutexes and the order of these operations. Incorrect order of unlocking the mutexes or the total failure to unlock the mutex can lead to the deadlock. An example of this is presented by the following application model (C_{AP2}):

$$T_{AP2} = (t_0, t_1, t_2),$$
$$U_{AP2} = (\{t_0\}, \{t_1, t_2\}, \{t_0\}),$$
$$R_{AP2} = \{(r1, 0), (r2, 0)\},$$
$$O_{AP2} = \{o_{0,1}, o_{0,2}, o_{1,1}, o_{1,2}, o_{1,3}, o_{2,1},$$
$$o_{2,2}, o_{2,3}\},$$

$$MU^2_{AP2} = \{\{o_{1,1}, o_{1,2}, o_{1,3}, o_{2,1}\},$$
$$\{o_{1,1}, o_{1,2}, o_{1,3}, o_{2,1}\}\}$$
$$F_{AP2} = \{(o_{0,1}, o_{0,2}), (o_{1,1}, o_{1,2}),$$
$$(o_{1,2}, r_1), (o_{1,2}, o_{1,3}), (o_{1,3}, r_2),$$
$$(o_{2,1}, o_{2,2}), (o_{2,2}, r_1), (o_{2,2}, o_{2,3}),$$
$$(o_{2,3}, r_2)\}.$$

Figure 4 shows the visualization of the application (C_{AP2}) in the proposed graphic notation (Fig. 4b) and the equivalent of this application expressed by PN (Fig. 4a), built in accordance with [7]. In the case of PN, mutexes $m_{1,1}$ and $m_{1,2}$ in Fig. 4a are assumed at the beginning of the branches responsible for the work of the threads. However, in Fig. 4a there is no information about unlocking the mutexes, as a result of which the mutexes remain active, and this leads to a situation where all operations occurring after locking are excluded. In the case of the proposed model, the mutexes are active only at the specified time u2. They include only those operations that are actually protected in the code by means of a lock. This allows obtaining information that operations $o_{2,2}$ and $o_{2,3}$ in t2 are not mutually exclusive with t1 thread operations, which leads to race condition. From the drawing (Fig. 4b) it is also possible to read the order in which the mutexes are locked and unlocked. The order of these operations is identical in both threads, which excludes the occurrence of a deadlock. Another difference of the proposed model in relation to the PN (and usually other methods based on PN) is the lack of ambiguity. In the case of PN, the ambiguity lies in the fact that for one multithreaded application it is possible to build many network models [1], which will contain redundant elements, e.g. p1, p2, p3 and p4 from Fig. 4a. The proposed model is not ambiguous because there are no excess elements.

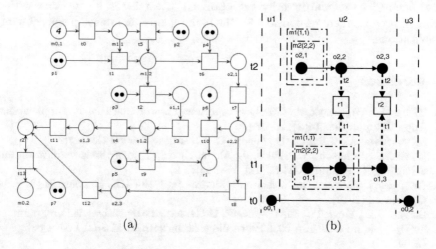

(a)   (b)

Fig. 4. Example of race condition not detected by PN.

6   Conclusion

The work describes a formal model and graphical representation that allow mapping multithreaded applications. The most characteristic feature of the developed model is the inclusion of time as one of the variables. This feature allows you to assess which of the threads work in parallel at the same time, and thus whether a situation is possible in which undesirable phenomena will occur. The second feature of the developed model is the division of operations and resources into separate entities and the determination of relations between them. Thanks to such a procedure, the graphic representation is characterized by transparency and there are no ambiguities in it, which allow to confuse these two elements. The third feature of the model are the functions ϕ and ψ, which value is determined by the order of locking and unlocking mutexes. This feature is very important, because the incorrect sequence of locking and unlocking them may lead to the deadlock. Thanks to, among others, these features of the model can detect race condition and determine which operations use parallel shared resources.

So far, research into the model has been theoretical and concerns only C's multithreaded applications that use the *pthread*. One of the challenges to be taken to perform further experiments and practical verification is to automate the model building process based on the program code, which should be a process reversible. Other challenges are related to the application of the model to programming languages based on the object-oriented programming paradigm and other libraries than *pthread*.

Further works on the development of the model predicts the aforementioned automation of the model construction and the extension of its structure with elements necessary to detect in the multithreaded applications the remaining untidy phenomena, i.e. deadlocks, atomicity violation and order violation. It is also planned to regularly verify the obtained research results together with a team of programmers responsible for the maintenance and development of many applications.

References

1. Giebas, D., Wojszczyk, R.: Graphical representations of multithreaded applications. Appl. Comput. Sci. **14**(2), 20–37 (2019)
2. Silberschatz, A., Galvin, P.B., Gagne, G.: Operating System Concepts
3. Clarke, E., Emerson, A., Sifakis, J.: Model checking: algorithmic verification and debugging. Commun. ACM **52**(11), 74–84 (2009)
4. Greniewski, M.J.: Network software validation. In: 17th KKIO Software Engineering Conference (2016)
5. Berger, E.D., Yang, T., Liu, T., Novark, G.: Grace: safe multithreaded programming for C/C++. In: OOPSLA 2009 Proceedings of the 24th ACM SIGPLAN Conference (2009)

6. Shan, L., Soyeon, P., Eunsoo, S., Yuanyuan, Z.: Learning from mistakes – a comprehensive study on real world concurrency bug characteristics. In: Proceedings of the 13th International Conference on Architectural Support for Programming Language and OS (2008)
7. Kavi, K.M., Moshtaghi, A., Chen, D.: Modeling multithreaded applications using Petri nets. Int. J. Parallel Program. **30**, 353–371 (2002)

Special Session on Theoretical Foundations and Mathematical Models in Computer Science, Artificial Intelligence and Big Data (TMM-CSAIBD)

Special Session on Theoretical Foundations and Mathematical Models in Computer Science, Artificial Intelligence and Big Data (TMM-CSAIBD)

Special session on Theoretical Foundations and Mathematical Models in Computer Science, Artificial Intelligence and Big Data (TMM-CSAIBD) is intended to bring together researchers working in the areas of theoretical and applied Mathematics in Computer science, Artificial Intelligence and Big Data to provide a high-quality forum for the dissemination and discussion of research results in these broad areas. TMM-CSAIBD has originated from the ongoing efforts for promoting research in these fields.

This special session held under aegis of 16th International Conference on Distributed Computing and Artificial Intelligence (DCAI) and will focus on the theoretical and applied mathematical model applied to new data science technologies. This special session is devoted to promoting the investigation of the latest research of mathematical models and foundations and their effective applications, to explore the latest innovations in guidelines, theories, models, ideas, technologies, applications and tools of Computer science, Artificial Intelligence and Big Data to assess the impact of the approach, and to facilitate technology transfer.

Organization

Organizing Committee

Ángel Martín del Rey	Universidad de Salamanca, Spain
Roberto Casado-Vara	Universidad de Salamanca, Spain
Fernando De la Prieta	Universidad de Salamanca, Spain

Program Committee

Zita Vale	Politécnico do Porto, Portugal
Ángel Martín del Rey	Universidad de Salamanca, Spain
Pastora Vega	University of Salamanca, Spain
G. Kumar Venayagamoorthy	Clemson University, SC, USA
Héctor Quintian	University of A Coruña, Spain
Javier Prieto	Universidad de Salamanca, Spain
Esteban Jove-Pérez	Universidad de la Coruña, Spain
Fernando De la Prieta	Universidad de Salamanca, Spain
Paulo Novais	Universidad do Minho, Portugal
Slobodan Petrovic	Norwegian University of Science and Technology, Norway
José Luis Calvo-Rolle	University of A Coruña, Spain
Roberto Casado-Vara	Universidad de Salamanca, Spain

Modeling the Spread of Malware
on Complex Networks

A. Martín del Rey[1(✉)], A. Queiruga Dios[1], G. Hernández[2],
and A. Bustos Tabernero[3]

[1] Institute of Fundamental Physics and Mathematics, University of Salamanca,
Salamanca, Spain
{delrey,queirugadios}@usal.es
[2] BISITE Research Group, University of Salamanca, Salamanca, Spain
guillehg@usal.es
[3] Faculty of Sciences, University of Salamanca, Salamanca, Spain
alvarob97@usal.es

Abstract. Currently, zero-day malware is a major problem as long as
these specimens are a serious cyber threat. Most of the efforts are focused
on designing efficient algorithms and methodologies to detect this type of
malware; unfortunately models to simulate its behavior are not well studied. The main goal of this work is to introduce a new individual-based
model to simulate zero-day malware propagation. It is a compartmental model where susceptible, infectious and attacked devices are considered. Its dynamics is governed by means of a cellular automaton whose
local functions rule the transitions between the states. The propagation
is briefly analyzed considering different initial conditions and network
topologies (complete networks, random networks, scale-free networks and
small-world networks), and interesting conclusions are derived.

Keywords: Malware · Propagation · Complex networks ·
Individual-based model · Cellular automata

1 Introduction

Malicious code (also known as malware) can be considered one of the major
threats to our digital society since the different types (computer viruses, trojans, computer worms, etc.) are the fundamental tools used in cyberattacks.
Special mention must be made to zero-day malware and its role in the Advanced
Persistent Threats (APTs for short).

APTs are sophisticated cyber-attacks [19] combining different technologies
and methodologies. They are characterized by the following: (1) they have a
precise and clear target, (2) they are long-term attacks constituted by several
phases of different nature (reconnaissance, incursion, discovery, capture and exfiltration), (3) they implement evasive techniques consisting on stealthy behavior
and adaptation to defenders' efforts, and (4) they are complex due to the use

E. Herrera-Viedma et al. (Eds.): DCAI 2019, AISC 1004, pp. 109–116, 2020.
https://doi.org/10.1007/978-3-030-23946-6_12

of different attack methods (zero-day malware, rootkits, etc.) Zero-day malware plays an important role in the implementation of APTs [18]. It is a specimen of malicious code which is characterized by exploiting unknowns (or non-patched) vulnerabilities [15].

The fight against malware is mainly based on the design of Machine Learning techniques and protocols to successfully detect it (see, for example [10,11]). Nevertheless, it is also important to simulate its propagation in an efficient way. In this sense several mathematical models have been proposed in the scientific literature [13]. Most of them are theoretical models of a continuous nature where the spreading environment is defined by a complete network where all devices are in contact with each other [2]. Due to their initial constraints, these models have limited practical application and this drawback has been tried to overcome by considering alternative topologies [1]. In recent years some proposals have appeared dealing with the simulation of malware propagation on complex networks: some of them are deterministic models [4,5,7,9,12], and other are stochastic (see, for example [6,8] and references therein). Although these models give rise to more realistic simulations, they do not take into account the individual characteristics of the devices. This is a very important issue and some attempts have been made using Artificial Intelligence techniques such as agent-based models [3,6] or cellular automata [14,17]. Nevertheless, as far as we know, no individual-based model considering zero-day malware has been proposed.

The main goal of this work is to introduce a novel model to simulate zero-day malware whose dynamics will be governed by means of a cellular automaton. It is an individual-based model where the devices are classified into three compartments: susceptible, infectious and attacked. Its dynamics considers some characteristics of zero-day malware: an infectious device becomes susceptible again if it is not the target of the attack. The model is flexible in the sense that different topological conditions are stated in order to determine the local interactions between the devices.

The rest of the paper is organized as follows: in Sect. 2 the definition of cellular automata is briefly introduced; the model for malware spreading is described in Sect. 3; in Sect. 4 the study of the effects of topology in the propagation is shown, and finally, the conclusions and future work are presented in Sect. 5.

2   Cellular Automata

A cellular automaton (CA for short) is a simple model of computation constituted by n identical memory units called cells [16]: $c_1, c_2, \ldots, c_n$, that are arranged following a certain topology defined by means of a complex network $\mathcal{G} = (\mathcal{V}, \mathcal{E})$. As a consequence, each cell of the cellular automaton stands for a vertex of $\mathcal{G}$: $c_i \in \mathcal{V}$ with $1 \leq i \leq n$, and the local interactions between these cells are modeled by means of the edges: there is a connection between c_i and c_j if $(c_i, c_j) \in \mathcal{E}$. Consequently, the neighborhood of the cell c_i is defined as the collection of its adjacent cells:

$$\mathcal{N}_i = \{v \in \mathcal{V} \text{ such that } (v, c_i) \in \mathcal{E}\} = \{c_{\alpha_1}, \ldots, c_{\alpha_i}\} \quad 1 \leq i \leq n. \quad (1)$$

Each cell c_i is endowed with a state at each step of time t: $s_i^t \in \mathcal{S}$, where $\mathcal{S}$ is a finite state set. This state changes accordingly to a local transition function f whose variables are the states of the main cell and the neighbor cells at the previous step of time:

$$s_i^{t+1} = f\left(s_i^t, s_{\alpha_1}^t, \ldots, s_{\alpha_i}^t\right) \in \mathcal{S}. \tag{2}$$

3   Description of the Model

The epidemiological model proposed in this work is a compartmental model where the population of devices is divided into three classes or compartments: susceptible, infectious and attacked. Susceptible devices are those that have not been infected by the malware (the device is free of the malicious code); the infectious devices are characterized because they have been reached by the malware but have not been attacked, and finally, attacked devices are those devices where the malware is carrying out its stealthy and malicious activity.

The dynamics of the propagation process is illustrated in Fig. 1. The coefficients involved are the following: the infection rate $0 \leq h_i \leq 1$ that rules the infection of a device free of malware, the targeted coefficient $0 \leq a_i \leq 1$ which defines the probability that the infectious device c_i will be effectively attacked, the recovery (from infectious) probability $0 \leq b_i << a_i \leq 1$, and the recovery (from attacked) coefficient $r_i (= 0 \text{ or } 1)$ that determines the auto-removing malware process.

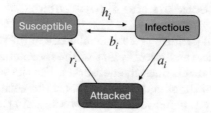

Fig. 1. Flow diagram representing the dynamics of the model.

Note that a susceptible device becomes infectious when the malware reaches it (and, in our model, this depends on both the infection rate and the number of infectious neighbor devices); the infectious device c_i becomes attacked with probability a_i or susceptible with probability $b_i << a_i$ (in this work it is supposed that the malware can remove itself if it does not find any neighbor host or the current host must not be attacked); finally, the attacked devices recover once the attacked period is finished. As a consequence, it is a SIAS model (Susceptible-Infectious-Attacked-Susceptible).

The propagation model is based on a cellular automaton whose main characteristics are the following:

- Each device represents a cell of the CA.
- The state set is formed by three states: susceptible (S), infectious (I), and attacked (A): $\mathcal{S} = \{S, I, A\}$.
- The local transition functions between the compartments are the following:
    - Transition from susceptible to infectious: if $s_i^t = S$ then $s_i^{t+1} = I$ with probability $h_i \cdot N_i^t$, where N_i^t is the number of infectious neighbor devices of c_i at t. Obviously, the device c_i remains susceptible $(s_i^{t+1} = S)$ with probability $1 - h_i \cdot N_i^t$.
    - Transition from infectious to attacked: if $s_i^t = I$ then $s_i^{t+1} = A$ with probability a_i.
    - Transition from infectious to susceptible: if $s_i^t = I$ then $s_i^{t+1} = S$ with probability b_i. Note that an infectious device c_i remains infectious at the next step of time with probability $1 - a_i - b_i$.
    - Transition from attacked to susceptible: finally, if $s_i^t = A$ then $s_i^{t+1} = S$ with probability r_i where:

$$r_i = \begin{cases} 1, & \text{if } t = t_i + \tau_i + 1 \\ 0, & \text{if } t_i + 1 \leq t \leq t_i + \tau_i \end{cases} \qquad (3)$$

where t_i is the step of time at which the device becomes attacked, and τ_i is the length of the attack period over the device c_i.

4   Topology Effects on the Propagation of Malware

In what follows we will perform several simulations of the model considering different initial conditions (numerical values of the parameters and characteristics of the topologies that determines the networks). For the sake of simplicity only illustrative examples of the simulations are shown (for each type of topology, the behaviors exhibited by the malware spreading are similar).

Suppose that the four epidemiological coefficients (infection rate, targeted rate, recovery —from infectious device— probability, and duration of the attack period) remain constant for every device: $h_i = \tilde{h}, a_i = \tilde{a}, b_i = \tilde{b}$ and $\tau_i = \tilde{\tau}$ with $1 \leq i \leq n$. Let's examine what happens if the topology is changed.

Assume that there are $n = 100$ devices in the network with only one infectious device at time step $t = 0$: the node with greater degree centrality (that is, with the largest number of neighbors). Set $\tilde{h} = 0.025, \tilde{a} = 0.25, \tilde{b} = 0.025$, and $\tilde{\tau} = 5$. Let $\mathcal{G}_1, \mathcal{G}_2, \mathcal{G}_3$, and $\mathcal{G}_4$ be a complete network, a random network –with edge probability $p = 0.5$–, a scale-free network –with 2 edges added at each step of the Barabasi-Albert algorithm–, and a small-world complex network –with rewiring probability $p = 0.1$ associated to the Watts-Strogatz algorithm–, respectively (see Fig. 2). Furthermore, some of the most important structural characteristics of these networks (density, diameter, radius and average path length) are shown in Table 1.

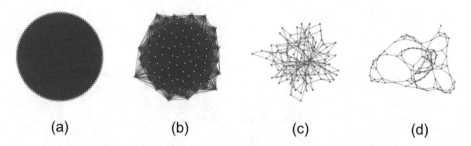

Fig. 2. (a) Complete network $\mathcal{G}_1$. (b) Random network $\mathcal{G}_2$. (c) Scale-free network $\mathcal{G}_3$. (d) Small-world network $\mathcal{G}_4$.

Table 1. Structural characteristics of the complex networks

Metric	$\mathcal{G}_1$	$\mathcal{G}_2$	$\mathcal{G}_3$	$\mathcal{G}_4$
Density	1	0.5103	0.03980	0.04040
Diameter	1	2	5	11
Radius	1	2	3	7
Average path length	1	1.490	3.044	5.042

Illustrative examples of the evolution of the different compartments (susceptible, infectious and attacked devices) are shown in Fig. 3 when the simulation of malware propagation during 90 units of time ($1 \leq t \leq 90$) is computed. Note that in the case of considering the complete topology the evolution of the different compartments exhibits an oscillatory behavior where the great majority of devices are infectious or attacked (see Fig. 3-(a)). The number of targeted devices (infectious and attacked) increases rapidly from the beginning and, in fact, at $t = 5$ the 100% of nodes are affected in some way by the malware. Furthermore, all nodes are infected or attacked during the simulation period.

A (qualitative) similar behavior is obtained when a random network $\mathcal{G}_2$ is considered (see Fig. 3-(b)). The propagation speed is lower but, as in the previous case, all nodes are infected and attacked by the malware. In this case the maximum is reached at $t = 10$ when the 98% of nodes are affected (46% of infectious devices, and 52% of attacked devices). The differences between the different device compartments are not as pronounced as for the complete case and the number of susceptible devices at every step of time is greater than in the precious case. Note that the graph density of $\mathcal{G}_2$ is approximately half of $\mathcal{G}_1$ and the diameter and radius of $\mathcal{G}_2$ is twice that of the complete network.

Very different from the previous dynamics are the evolutions of malware spreading obtained when scale-free or small-world networks are assumed (see Fig. 3-(c) and (d)). In these two cases the malware specimen barely manages to reach a neighbor node of the "patient zero". In fact, when $\mathcal{G}_3$ is considered only one effective contagion occurs; both devices were attacked and after that, the epidemic outbreak dies out. On the other hand, when the small-world network

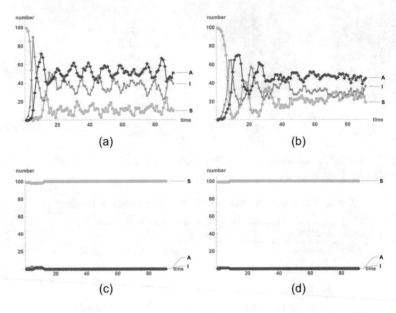

Fig. 3. Global evolution of malware on: (a) a complete network $\mathcal{G}_1$, (b) a random network $\mathcal{G}_2$, (c) a scale-free network $\mathcal{G}_3$, and (d) a small-world network $\mathcal{G}_4$.

$\mathcal{G}_4$ is considered, the malware fails to spread and only one targeted device exists (initially it was infectious and at time step $t = 3$ becomes attacked). This fact should not surprise us if we take into account that the densities of $\mathcal{G}_3$ and $\mathcal{G}_4$ are 0.04 and the average path lengths are much greater than those of $\mathcal{G}_1$ and $\mathcal{G}_2$.

5   Conclusions

In this work a novel computational model to simulate zero-day malware spreading is introduced and analyzed, where the environment is defined by means of a complete network, a typical random network, a scale-free network, and a small-world network. The model is compartmental considering susceptible, infectious and attacked devices. Its dynamics is characterized by the following:

(1) Two classes of targeted devices are considered: infectious (they serve as transmission vectors but they are not the target of the attack) and attacked (they are the main objectives of the attack).
(2) Infectious devices can immediately recover the susceptible status if they are not attacked.
(3) The duration of the malware payload is constant and, consequently, the attacked devices becomes susceptible again when the attacking period finishes.
(4) Only the stealthy behavior of the malware is taken into account. This is reflected when considering the local transition from infectious to susceptible and the short-term attack period.

Several simulations have been performed and only illustrative examples are shown in this paper. From a simple analysis of the same, the following conclusions follow:

- Some structural characteristics of the networks (density, diameter, radius, and average path length) have a direct influence on the propagation process. The malware spreading can be accelerated or decelerated depending on the value of these parameters.
- The type of topology also influences the propagation: complete or random complex networks (with probability greater that 0.5) exhibit a similar evolution: the initial outbreak becomes an epidemic process converging to an oscillatory behavior. On the other hand, scale-free and small-world complex networks slow down the propagation; in fact, the evolution tends to disease-free steady states.

Future work aimed at considering additional zero-day malware features in the design of the cellular automata (states and local transition functions). Furthermore, alternative scenarios must be analyzed in detail considering different choices of "patient zero" based on alternative centrality coefficients (clustering coefficient, betweenness coefficient, etc.) and different measures of the propagation speed.

Acknowledgements. This research has been partially supported by Ministerio de Ciencia, Innovación y Universidades (MCIU, Spain), Agencia Estatal de Investigación (AEI, Spain), and Fondo Europeo de Desarrollo Regional (FEDER, UE) under project with reference TIN2017-84844-C2-2-R (MAGERAN) and the project with reference SA054G18 supported by Consejería de Educación (Junta de Castilla y León, Spain).
    A. Bustos Tabernero thanks Ministerio de Educación y Formación Profesional (Spain) for his departmental collaboration grant in the Department of Applied Mathematics (University of Salamanca, Spain).

References

1. Fu, X., Small, M., Chen, G.: Propagation Dynamics on Complex Networks. Wiley, Hoboken (2014)
2. Hernández Guillén, J.D., Martín del Rey, A.: Modeling malware propagation using a carrier compartment. Commun. Nonlinear Sci. Numer. Simul. **56**, 217–226 (2018)
3. Hosseini, S., Azgomi, M.A., Torkaman, A.R.: Agent-based simulation of the dynamics of malware propagation in scale-free networks. Simulation **92**, 709–722 (2016)
4. Hosseini, S., Azgomi, M.A.: A model for malware propagation in scale-free networks based on rumor spreading process. Comput. Netw. **108**, 97–107 (2017)
5. Hosseini, S., Azgomi, M.A.: The dynamics of a SEIRS-QV malware propagation model in heterogeneous networks. Phys. A **512**, 803–817 (2018)
6. Hu, P., Ding, L., Hadzibeganovic, T.: Individual-based optimal weight adaptation for heterogeneous epidemic spreading networks. Commun. Nonlinear Sci. Numer. Simul. **63**, 339–355 (2018)
7. Jackson, J.T., Creese, S.: Virus propagation in heterogeneous bluetooth networks with human behaviors. IEEE Trans. Dependable Secur. Comput. **9**, 930–943 (2012)

8. Karyotis, V., Khouzani, M.H.R.: Malware Diffusion Models for Modern Complex Networks. Morgan Kaufmann-Elsevier, Cambridge (2016)
9. Karyotis, V., Papavassiliou, S.: Macroscopic malware propagation dynamics for complex networks with churm. IEEE Commun. Lett. **19**, 577–580 (2015)
10. Kim, J.-Y., Bu, S.-J., Cho, S.-B.: Zero-day malware detection using transferred generative adversarial networks based on deep autoencoders. Inform. Sci. **460**, 83–102 (2018)
11. Kim, T., Kang, B., Rho, M., Sezer, S., Im, E.G.: A multimodal deep learning method for Android malware detection using various features. IEEE Trans. Inf. Forensic Secur. **14**, 773–788 (2019)
12. Liu, W., Zhong, S.: A novel dynamic model for web malware spreading over scale-free networks. Phys. A **505**, 848–863 (2018)
13. Martín del Rey, A.: Mathematical modeling of the propagation of malware: a review. Secur. Commun. Netw. **8**, 2561–2579 (2015)
14. Martín del Rey, A., Rodríguez Sánchez, G.: A discrete mathematical model to simulate malware spreading. Int. J. Mod. Phys. C **23**, 1–16 (2012). Article number 1250064
15. Rudd, E.M., Rozsa, A., Günter, M., Boult, T.E.: A survey of stealth malware attacks, mitigation measures, and steps toward autonomous open world solutions. IEEE Commun. Surv. Tutor. **19**, 1145–1172 (2017)
16. Sarkar, P.: A brief history of cellular automata. ACM Comput. Surv. **32**, 80–107 (2000)
17. Thomson, B., Morris-King, J.: An agent-based modeling framework for cybersecurity in mobile tactical networks. J. Def. Model. Simulat. **15**, 204–218 (2018)
18. Tounsi, W., Rais, H.: A survey on technical threat intelligence in the age of sophisticated cyber attacks. Comput. Secur. **72**, 212–233 (2018)
19. Winkler, I., Treu Gomes, A.: Advanced Persistent Security. A Cyberwarfare Approach to Implementing Adaptive Enterprise Protection, Detection, and Reaction Strategies. Syngress-Elsevier, Cambridge (2017)

Application of the Bayesian Model in Expert Systems

Francisco João Pinto[⊠]

Department of Computer Engineering, Faculty of Engineering,
University Agostinho Neto, University Campus of the Camama, S/N,
Luanda, Angola
fjoaopinto@yahoo.es

Abstract. In this paper, we studied the basic elements of the Bayesian model, the elementary application of the rule of Bayes in the field of medical diagnosis, the Bayesian model in expert systems and finally the advantages and inconveniences of the Bayesian model. In fact that, the Bayesian methods been based on the theory of probabilities, nowadays they are popular in the artificial intelligence world as methods to treat the uncertainty. However, during many years it was thought that these methods were impracticable in real problems due to great amount of probabilities that is necessary to obtain in order to build a base of knowledge but this situation changed with the arrival of the networks of belief or Bayesian networks because the use of these networks allows to build bases of knowledge consistent probabilistic. The result of investigation demonstrate that the main advantage of the Bayesian methods resides in that they are strongly based in the theory of the probability, however, its main difficulty is in the great quantity of probabilities that is necessary to obtain in order to build a base of knowledge.

Keywords: Bayesian model · Expert system · Artificial intelligence

1 Introduction

Expert systems were introduced around 1965 [8] by the Stanford Heuristic Programming Project led by Edward Feigenbaum, who is sometimes termed the "father of expert systems"; other key early contributors were Bruce Buchanan and Randall Davis. The Stanford researchers tried to identify domains where expertise was highly valued and complex, such as diagnosing infectious diseases (Mycin) and identifying unknown organic molecules (Dendral). The idea that "intelligent systems derive their power from the knowledge they possess rather than from the specific formalisms and inference schemes they use" [8] – as Feigenbaum said – was at the time a significant step forward, since the past research had been focused on heuristic computational methods, culminating in attempts to develop very general-purpose problem solvers (foremostly the conjunct work of Allen Newell and Herbert Simon) [8]. Expert systems became some of the first truly successful forms of artificial intelligence (AI) software [9]. Research on expert systems was also active in France. While in the US the focus tended to be on rule-based systems, first on systems hard coded on top of LISP programming

E. Herrera-Viedma et al. (Eds.): DCAI 2019, AISC 1004, pp. 117–124, 2020.
https://doi.org/10.1007/978-3-030-23946-6_13

environments and then on expert system shells developed by vendors such as *Intellicorp*, in France research focused more on systems developed in Prolog. The advantage of expert system shells was that they were somewhat easier for nonprogrammers to use. The advantage of Prolog environments was that they weren't focused only on *if-then* rules; Prolog environments provided a much better realization of a complete First Order Logic environment [10].

In the 1980s, expert systems proliferated. Universities offered expert system courses and two thirds of the Fortune 500 companies applied the technology in daily business activities [11]. Interest was international with the Fifth Generation Computer Systems project in Japan and increased research funding in Europe. In 1981, the first IBM PC, with the PC DOS operating system, was introduced. The imbalance between the high affordability of the relatively powerful chips in the PC, compared to the much more expensive cost of processing power in the mainframes that dominated the corporate IT world at the time, created a new type of architecture for corporate computing, termed the client-server model [11]. Calculations and reasoning could be performed at a fraction of the price of a mainframe using a PC. The first expert system to be used in a design capacity for a large-scale product was the SID (Synthesis of Integral Design) software program, developed in 1982. Written in LISP, SID generated 93% of the VAX 9000 CPU logic gates [11]. Input to the software was a set of rules created by several expert logic designers.

In the 1990s and beyond, the term *expert system* and the idea of a standalone AI system mostly dropped from the IT lexicon. There are two interpretations of this. One is that "expert systems failed": the IT world moved on because expert systems didn't deliver on their over hyped promise [12]. The other is the mirror opposite, that expert systems were simply victims of their success: as IT professionals grasped concepts such as rule engines, such tools migrated from being standalone tools for developing special purpose *expert* systems, to being one of many standard tools [12]. Many of the leading major business application suite vendors (such as SAP, Siebel, and Oracle) integrated expert system abilities into their suite of products as a way of specifying business logic – rule engines are no longer simply for defining the rules an expert would use but for any type of complex, volatile, and critical business logic; they often go hand in hand with business process automation and integration environments [12].

2   Basic Elements of the Bayesian Model

First of all, we will examine some fundamental concepts of the theory of the probabilities. Be A an event, the collection of all the possible events elementary M is known like space of events. The probability of an event A is denoted with p(A), and all function of probability p must satisfy three axioms [1]:

(1) The probability of any elementary event A is not negative, that is to say, $\forall A \in \Omega : P(A) \geq 0$.
(2) The probability of the space of events is one, that is to say, $p(\Omega) = 1$.

(3) If K events $A_1, A_2, \ldots, A_k$ are excluding (that is to say, they cannot occur simultaneously), then the probability that at least one of those events occurs it is the sum of the individual probabilities:

$$p(A_1 \cup A_2 \cup \ldots \cup A_k) = \sum_{i=1}^{k} p(A_i)$$

From these axioms we can deduce others. Like this, of the axioms 1 and 2, we can obtain the following result: $\forall A \in \Omega : 0 \leq p(A) \leq 1$

This equation shows that the probability of any event is between zero and one. By definition if $p(A) = 0$ the event A never occurs, and when $p(A) = 1$ the event A would occur.

The complementary of A (*represented as* $\neg A$) it contains the collection of all the elementary events in Ω except A. As A and $\neg A$ are mutually excluding and $A \cup \neg A = \Omega$ by the axiom 3, we can obtain the following result:

$p(A) + p(\neg A) = p(A \cup \neg A) = p(\Omega) = 1$. Rewrite this equation like $p(\neg A) = 1 - p(A)$, we obtain a simple method to calculate $p(\neg A)$ from $p(A)$. If we suppose that B is an event, the probability that A occurs, knowing that B occurs, represented by $p(A/B)$, it is known as the conditional probability of A given B. This probability is defined as:

$$p(A/B) = \frac{p(A \cap B)}{p(B)} \tag{1}$$

From this equation, it is simple to obtain the rule of Bayes, base of the Bayesian model, in the following way: the probability of B given A is p(B/A) = p(B)p(A) like the combined probability it is commutative then:

$p(A \cap B) = p(B \cap A)$ With that we can write: $p(B \cap A) = p(A \cap B) = p(B/A)p(A)$, substituting this expression in the definition of conditional probability of A given B we can obtain the rule of Bayes:

$$p(A/B) = \frac{p(B/A)p(A)}{p(B)} \tag{2}$$

The basic equation of Bayes can be generalized considering that:

$$p(B) = p(B \cap A) \cup p(B \cap \neg A)$$
$$= p(B \cap A) + p(B \cap \neg A)$$
$$= p(B/A)p(A) + p(B/\neg A)p(\neg A)$$

Substituting this expression of p(B) in the rule of Bayes we can obtain the following equation:

$$p(A/B) = \frac{p(B/A)p(A)}{p(B/A)p(A) + p(B/\neg A)p(\neg A)} \tag{3}$$

That it allows us to avoid to value the probability a priori p(B) in exchange for valuing the conditional probability $p(B/\neg A)$. This expression can be generalized as:

$$p(A_0/B) = \frac{p(B/A_0)p(A_0)}{\sum_i p(B/A_i)P(A_i)} \tag{4}$$

3   Elementary Application of the Rule of Bayes

Seemingly, the basic rule of Bayes doesn't seem to be of a lot of utility. Three terms are needed (a conditional probability and two unconditional probabilities) only to be able to calculate a probability. However, the regla of Bayes is useful in practice because in many cases present good estimates of probabilities for these three numbers and therefore it are necessary to calculate the fourth number.

A simple example of the application of the basic rule of Bayes can be in the field of the medical diagnosis. Let us suppose that we know the probability a priori of a certain illness E. We also know a priori the probability of appearance of a certain symptom S and the conditional probability of S given E. In such case, if a patient appears during the consultation a with symptom S, the rule of Bayes allows us to obtain the probability that the patient suffers of the illness E [6]:

$$p(E/S) = \frac{p(S/E)p(E)}{p(S)}$$

However, a question that probably has arisen to all those that analyze the rule of Bayes **by** the first time is: why the probability p (E/S) is necessary to calculate and the probability p(S/E) it gets ready beforehand? Why is not obtained the probability P(E/S) beforehand? The answer to these questions is that, the knowledge obtained by diagnosis is weaker than the causal knowledge. The probability P(S/E) is causal information, that is to say, it is the probability that a certain illness causes a certain symptom. These probabilities are usually quite well known and they are not affected by external conditions.

However, the probability of having a certain illness knowing that one has a certain symptom, p(E/S), it is not so easy of calculating because it depends on external conditions (for example, an epidemic of the illness E). The rule of Bayes allows us to obtain the probability p(E/S) through a conditional probability that can be easy to fix, p (S/E), and two probabilities a priori, p(E) and p(S) that reflect the relationship of the domain with the illness and the symptom [1].

4 The Bayesian Model in Expert Systems

The rule of Bayes expressed in the Eq. (4) is the base for the use of the theory of the probability in the handling of the uncertainty. The task of an expert system is to deduce what hypothesis H is certain given a certain group of evidences E. The Eq. (4), rewritten below based on H and E, it represents the case in which exist several hypotheses with a single evidence.

$$p(H_0/E) = \frac{p(E/H_0)p(H_0)}{\sum_i p(E/H_i)p(H_i)} \tag{5}$$

As we can see, the rule of Bayes allows us to obtain the probability of a given hypothesis, based on the conditional probabilities of observing a given evidence, a hypothesis, p(E/H), and the probabilities a priori of all the hypotheses, p(H). These probabilities will reside in the base of knowledge of the expert system. However, the Eq. (5) should be generalized to contemplate the probability of appearance of multiple evidences [2]:

$$p(H_0/E_1, E_2, \ldots, E_n) = \frac{p(E_1, E_2, \ldots, E_n/H_0)p(H_0)}{\sum_i p(E_1, E_2, \ldots, E_n/H_i)p(H_0)} \tag{6}$$

The problem with this new version of the rule of Bayes is the combinatorial explosion that it is produced while the number of evidences grows. Therefore, the number of conditional probabilities that we could store is equal to $(n^0 H) \times 2^{(n^0 E)}$ always keeping in mind that the evidences can only take two values, true or false. Therefore, for a simple problem that 10 evidences indicate to 3 possible hypotheses, we found that the number of conditional probabilities that we have to store is of $3 \times 2^{10} = 3 \times 1024 = 3072$ a clearly high number for a clearly simple problem. The solution to the problem of the combinatorial explosion consists on supposing that the evidences are independent conditionally. Two evidences E_1 and E_2 are independent conditionally when its combined probability, given a certain hypothesis H, is equal to the product of the conditional probabilities of each given event H, or it is the same thing $p(E_1, E_2/H) = p(E_1/H)p(E_2/H)$. Like this, we can rewritten the Eq. (6) as [2]:

$$p(H_0/E_1, E_2, \ldots, E_n) = \frac{p(E_1/H_0)p(E_2/H_0)\ldots p(E_n/H_0)p(H_0)}{\sum_i p(E_1/H_i)p(E_2/H_i)\ldots p(E_n/H_i)p(H_i)} \tag{7}$$

Supposing a conditional independence, the number of conditional probabilities to store becomes:$(n^0 H) \times (n^0 E)$. For example, with 10 evidences and 3 hypothesis the result is 30. A Clearly inferior number to the obtained result of 3072 supposing dependence among the hypotheses.

To illustrate the application of the rule of Bayes, we can use the following example: Let us suppose that we have an exhaustive group formed by three mutually excluding hypothesis $(H_1, H_2 \text{ and } H_3)$ and two independent evidences $(E_1 \text{ and } E_2)$. The conditional probabilities a priori they are shown in the Table 1:

Table 1. The conditional probabilities

	i = 1	i = 2	i = 3
$p(H_i)$	0.5	0.3	0.2
$p(E_1/H_i)$	0.4	0.8	0.3
$p(E_2/H_i)$	0.7	0.9	0.0

A priori, the hypothesis that more probability has of being certain is H_1. However, as they go appearing evidences, the belief in the hypotheses will increase or it will diminish consequently. For example, imagine that we observe the evidence E_1, if we calculate the probability a posteriori of the hypotheses basing us on the Eq. (7), we obtain the following values:

$$p(H_1/E_1) = \frac{(0.4 \times 0.5)}{(0.4 \times 0.5) + (0.8 \times 0.3) + (0.3 \times 0.2)} = 0.4$$

$$p(H_2/E_1) = \frac{(0.8 \times 0.3)}{(0.4 \times 0.5) + (0.8 \times 0.3) + (0.3 \times 0.2)} = 0.48$$

$$p(H_2/E_1) = \frac{(0.3 \times 0.2)}{(0.4 \times 0.5) + (0.8 \times 0.3) + (0.3 \times 0.2)} = 0.12$$

As we can see, after the appearance of the evidence E_1, the belief in the hypotheses $(H_1 \text{ and } H_3)$ has diminished while the belief in H_2 has increased. If we observe the evidence E_2, we calculated the probabilities the subsequent again:

$$p(H_1/E_1, E_2) = \frac{(0.4 \times 0.7 \times 0.5)}{(0.4 \times 0.7 \times 0.5) + (0.8 \times 0.9 \times 0.3) + (0.3 \times 0.0 \times 0.2)} = 0.39$$

$$p(H_2/E_1, E_2) = \frac{(0.8 \times 0.9 \times 0.3)}{(0.4 \times 0.7 \times 0.5) + (0.8 \times 0.9 \times 0.3) + (0.3 \times 0.0 \times 0.2)} = 0.61$$

$$p(H_3/E_1, E_2) = \frac{(0.3 \times 0.0 \times 0.2)}{(0.4 \times 0.7 \times 0.5) + (0.8 \times 0.9 \times 0.3) + (0.3 \times 0.0 \times 0.2)} = 0.00$$

Therefore, after the appearance of the evidences E_1 and E_2 there are only two relevant hypotheses H_1 and H_2, being H_2 the most probable.

5   Advantages and Inconveniences of the Bayesian Model

The main advantage of the Bayesian methods resides in that they are strongly based in the theory of the probability, however, its main difficulty is in the great quantity of probabilities that is necessary to obtain in order to build a base of knowledge. Thus, even supposing mutually excluding hypothesis, conditionally independent evidence and restricted variables to two values (true and false); if a medical diagnosis implies to 50 possible illnesses based on 100 symptoms, it is necessary to have 5000 conditional probabilities and 50 probabilities a priori.

Unfortunately, the supposition of conditional independence is rarely valid and the supposition of exclusivity mutual and exhaustiveness of the hypotheses is usually false; being the most frequents the appearance of concurrent hypothesis and superimposed. The Bayesians methods don't allow a clear explanation of their conclusions and they allow that same evidence supports, at the same time, a hypothesis and its negation [7].

An approach to solve these problems is the belief networks. A belief network is a special type of influence diagram in which the nodes represent random variables [4]. Demonstrated that the use of belief networks allows building bases of knowledge solid probabilistic, without imposing unnecessary assumptions of conditional independence. These networks also assure that the evidence in favor of a hypothesis won't be built partially by support of its negation, and that solid explanations can be obtained by means of the tracking of the beliefs until the initial points of the network.

6   Conclusions and Future Work

The Bayesians methods don't allow a clear explanation of their conclusions and they allow that a same evidence to support, at the same time, an hypothesis and its negation. An approach to solve these problems is the belief networks. A belief network is a special type of influence diagram in which the nodes represent random variables. It was demonstrated that the use of belief networks allows building bases of knowledge solid probabilistic, without imposing unnecessary assumptions of conditional independence. For further work, we will study in-depth the belief networks for the resolution of the problem of the great amount of probabilities that is necessary to obtain to build a base of knowledge using Bayesian model. The use of belief networks allows building bases of knowledge solid probabilistic, without imposing unnecessary assumptions of conditional independence.

References

1. Bonillo, V.M., Betanzos, A.A., Canosa, M.C., Berdiñas, B.G., Rey, E.M.: Fundamentos de Inteligencia Artificial, Capítulo 4, páginas 87–93, 97–98, 111–115, Universidad de la Coruña-España, A Coruña (2000)
2. Castillo, E., Guitiérrrez, J.M., Hadi, A.S.: Systems experts y modelos de redes probabilisticas, Monografias de la Academia de Ingeniéria (1996)
3. McGrayne, S.B.: The Theory That Would Not Die: How Bayes' Rule Cracked the Enigma Code Hunted Down Russian Submarines & Emerged Triumphant from Two Centuries of Controversy. Yale University Press, New Haven (2011). ISBN 978-0-300-18822-6
4. Pearl, J.: Probabilistic Reasoning in Intelligent Systems: Networks of Pluasible Inference. Morgan Kauffman Publishers, Burlington (1998)
5. Russel, S.J., Noving, P.: Inteligencia Artificial: Un Enfoque Moderno. Prentice-Hall Hispano-americano, Madrid (1996)

6. Martel, P.J., Vegas, F.J.D.: Probabilidad y estadística en medicina. Diaz de Santos, Madrid (1997)
7. Jensen, F.V.: An Introduction to Bayesian Networks. UCL Press, London (1996)
8. Hayes-Roth, F., Waterman, D., Lenat, D.: Bulding Expert Systems, pp. 6–7. Addilson-Wesley, Boston (1983). ISBN 978-0-201-10686-2
9. Russell, S., Norvig, P.: Artificial intelligence: A Modern Approach (PDF), pp. 22–23. Simon & Schuster, New York (1995). 978-0-13-1038005-9
10. Durkin, J.: Expert Systems: Catalog of Applications. Intelligent Computer Systems, Inc, Akron (1993)
11. Leondes, C.T.: In: Expert Systems: The Technology of Knowledge, Management and Decisión the 21st Century, pp. 1–22 (2002). ISBN 978-0-12-443889-4
12. Zhao, K., Ying, S., Zhang, L., Hu, L.: Archiving Business Process and Business Rules Integration using SPL. Future Information Technology (2010). ISBN 978-1-42449087-5

Formal Representations of the Knowledge

Francisco João Pinto[(✉)]

Department of Computer Engineering, Faculty of Engineering,
University Agostinho Neto, University Campus of the Camama, S/N,
Luanda, Angola
fjoaopinto@yahoo.es

Abstract. This paper, describes the problem of the representation of the knowledge. After a brief discussion on the phase of coding - decoding are mentioned some of the characteristics and conditions that must assemble all scheme of representation of the knowledge in artificial intelligence (AI). The desire of to formalize the representation of the knowledge leads us to the introduction of the logic of propositions. Nevertheless, the limitations of the logic of propositions, which allows us neither represent several examples of the same entity, nor to quantify; it leads us to preferring the logic of predicates as scheme of formal representation. The results of the investigation demonstrate that the representation of the knowledge by logic of propositions is not versatile and thus come up the need of using, always from the formal optics, the logic of predicates.

Keywords: Formal representations · Knowledge · Artificial intelligence

1 Introduction

Knowledge representation and reasoning (KR) is the field of artificial intelligence (AI) dedicated to representing information about the world in a form that a computer system can utilize to solve complex tasks such as diagnosing a medical condition or having a dialog in a natural language. Knowledge representation incorporates findings from psychology about how humans solve problems and represent knowledge in order to design formalisms that will make complex systems easier to design and build. Knowledge representation and reasoning also incorporates findings from logic to automate various kinds of reasoning, such as the application of rules or the relations of sets and subsets. Examples of knowledge representation formalisms include semantic network, systems architecture, frames, rules, and ontology. Examples of automated reasoning engines include inference engines, theorem tester, and classifiers. The KR conference series was established to share ideas and progress on this challenging field [1].

The earliest work in computerized knowledge representation was focused on general problem solvers such as the General Problem Solver (GPS) system developed by Allen Newell and Herbert A. Simon in 1959. These systems featured data structures for planning and decomposition. The system would begin with a goal. It would then decompose that goal into sub-goals and then set out to construct strategies that could

© Springer Nature Switzerland AG 2020
E. Herrera-Viedma et al. (Eds.): DCAI 2019, AISC 1004, pp. 125–132, 2020.
https://doi.org/10.1007/978-3-030-23946-6_14

accomplish each sub goal. It was the failure of these efforts that led to the cognitive revolution in psychology and to the phase of AI focused on knowledge representation that resulted in expert systems in the 1970s and 80s, production systems, frame languages, etc. Rather than general problem solvers, AI changed its focus to expert systems that could match human competence on a specific task, such as medical diagnosis. Expert systems gave us the terminology still in use today where AI systems are divided into a Knowledge Base with facts about the world and rules and an inference engine that applies the rules to the knowledge base in order to answer questions and solve problems. In these early systems the knowledge base tended to be a fairly flat structure, essentially assertions about the values of variables used by the rules [4]. In addition to expert systems, other researchers developed the concept of frame based languages in the mid 1980s. A frame is similar to an object class: It is an abstract description of a category describing things in the world, problems, and potential solutions. Frames were originally used on systems geared toward human interaction, e.g. understanding natural language and the social settings in which various default expectations such as ordering food in a restaurant narrow the search space and allow the system to choose appropriate responses to dynamic situations. It wasn't long before the frame communities and the rule-based researchers realized that there was synergy between their approaches. Frames were good for representing the real world, described as classes, subclasses, slots (data values) with various constraints on possible values. Rules were good for representing and utilizing complex logic such as the process to make a medical diagnosis. Integrated systems were developed that combined Frames and Rules [6].

Knowledge-representation is the field of artificial intelligence that focuses on designing computer representations that capture information about the world that can be used to solve complex problems. The justification for knowledge representation is that conventional procedural code is not the best formalism to use to solve complex problems. Knowledge representation makes complex software easier to define and maintain than procedural code and can be used in systems. For example, talking to experts in terms of business rules rather than code lessens the semantic gap between users and developers and makes development of complex systems more practical. Knowledge representation goes hand in hand with automated reasoning because one of the main purposes of explicitly representing knowledge is to be able to reason about that knowledge, to make inferences, assert new knowledge, etc. Virtually all knowledge representation languages have a reasoning or inference engine as part of the system [4].

2   General Aspects of the Representation of the Knowledge

The formal logic allows the utilization of procedures of resolution that make possible the reasoning facts. In any application domain we will always find two types of different entities [2]:

- The facts, or truths of the domain
- The representations of the facts

The representations of the facts are the internal structures that the artificial intelligence programs manipulate, and corresponding with the truths of the domain. To manipulate computationally the facts, we need to define a group of procedures that transform such facts into representations. Once executed our program of AI, we need new procedures that transform the internal representations into comprehensible facts for us. The global process configures what is denominated coding-decoding phase (Fig. 1).

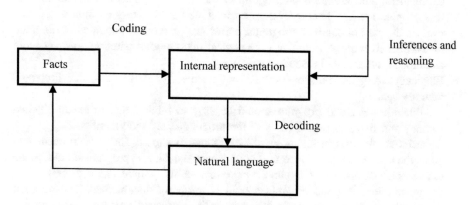

Fig. 1. The code-decoding cycle

The schemes of representation of the knowledge can be classified in one of the following categories:

- Declarative methods
- Procedural methods

The declarative schemes the knowledge represents like a static collection of facts for whose manipulation defines a generic group and restricted of procedures. The schemes of this type present some advantages: The truths of the domain are stored only by one time. Besides, it is easy to increase and incorporate new knowledge without modifying neither alter the already existent. In the procedural diagrams the greater part of the knowledge represents as procedures, which confers the scheme of representation a dynamic character. Also the procedural schemes present advantages [2]:

- When giving priority to the procedures do greater emphasis in the capacities of inferences of the system
- They allow to explore distinct models and technical of reasoning
- They allow to work with fault of information and with data of probabilistic character
- They incorporate of natural form knowledge of heuristic type.

We should also know that a fundamental question, related to good programming is: What is a good program? Answering this question is not trivial as there are several criteria for judging how good a program is. Generally accepted criteria include the following [5]:

- Transparency, readability: A good program should be easy to read and easy to understand. It should not be more complicated than necessary. Clever programming tricks that obscure the meaning of the program and its layout help its readability.
- Efficiency: A good program should not needlessly waste computer time and memory space.
- Correctness: Above all, a good program should be correct. That is, i should do what it is supposed to do. This may seem a trivial, self explanatory requirement. However, in the case of complex programs, correctness is often not attained. A common mistake when writing programs is to neglect this obvious criterion and pay more attention to other criteria, such as efficiency.
- Efficiency: A good program should not needlessly waste computer time and memory space.
- Modifiability: A good program should be easy to modify and to extend. Transparency and modular organization of the programa help modifiability.
- Robustness: A good program should be robust. It should not crash immediately when the user enters some incorrect or unexpected data. The program should, in the case of such errors, stay alive and behave reasonably (should report errors).
- Documentation: A good program should be properly documented. The minimal documentation is the program's listing including sufficient program comments.

The importance of particular criteria depends on the problems and on the circumstances in which the program is written, and on the environment in which it is used. There is no doubt that correctness has the highest priority. The issues of transparency, modifiability, robustness and documentation are usually given, at least, as much priority as the issue of efficiency [7].

There are some general guidelines for practically archiving the above criteria. One important rule is to first think about the problem to be solved, and only then to start writing the actual code in the programming language used. Once we have developed a good understanding of the problem and the whole solution is well thought thought, the actual coding will be fast and easy, and there is a good chance that we will soon get a correct program [7].

A common mistake is to start writing the code even before the full definition of the problem has been understood. A fundamental reason why early coding is bad practise is that the thinking about the problem and ideas for a solution should be done in terms that are most relevant to the problem. These terms are usually far from the syntax of programming language used, and they may include natural language statements and pictorial representation of ideas [5, 7].

According to the principle of stepwise refinement, the final program developed through a sequence of transformations, or 'refinements', of the solution. We start with the first, top-level solution and the proceed through a sequence of solutions; these are all equivalent, but each solution in the sequence is expressed in more detail. In each refinement step, concepts used in previous formulations are elaborated to greater detail and their representation gets closer to the programming language. It should be realized that refinement applies both to procedure definitions and to data structures. In the initial stages we normally work with more abstract, bulky units of information whose structure is refined later [7].

In the case of Prolog we may talk about the stepwise refinement of relations. If the nature of the problem suggests thinking in algorithmic terms, then we can also talk refinement of algorithms, adopting the procedural point of view on Prolog.

In order to properly refine a solution at some level of detail, and to introduce useful concepts at the next lower level, we need ideas. Therefore programming is creative, especially so for beginners. With experience, programming gradually becomes less of an art and more of a craft. But, nevertheless, a major question is: How do we get ideas? Most ideas come from experience, from similar problems whose solutions we know. If we do not know a direct programming solution, another similar problems could be helpful. Another source of ideas is everyday life. For example, if the problem is to write a program to sort a list of ítems we may get an idea from considering the question. General principles of good programming outlined in this section are also known as the ingredients of 'structured programming', and they basically apply to Prolog as well [5].

3 Propositional Logic and Predicates Logic

Both the logics belong to the formal logic. As diagram of representation of the knowledge, the formal logic allows to derive knowledge starting from knowledge that already exists through deductive processes. Thus, in formal logic an asseveration is certain if we can demonstrate that one can deduce from other certain asseverations. The simplest in the formal logics is the logic of propositions, in which the facts of the real world are represented as logical propositions that they are well defined formulas or they are well formed formulas [3].

For example in the logic of propositions the declaration **"Francisco is a man"** can be denoted by any letter, in this case, we denoted by **P: Francisco is a man**. For construction, the logic of propositions, presents several problems. Thus, how we can represent efficiently several examples of a same entity? how can we solve the problem of the quantification?

The representation of the knowledge by means of logic of propositions is not versatile. It is important to always use from the formal optics, the logic of predicates. In logic of predicates the knowledge is represented as logical declarations that they are well defined formulas. Thus, we pass of structures of the type **P: Francisco is a man** to a structure of the type: **MAN (FRANCISCO)** or **IS_A (FRANCISCO, MAN)**.

The basic components of an diagram of representation of the knowledge based on predicate logic, are: Alphabet, formal language, axioms or groups of enunciated basic and inference rules.

The axioms, describe fragments of knowledge and the rules of inferences are applied to the axioms to deduce new true enunciated. In any formal language, the alphabet is a set of symbols that allows us to build the enunciated. In logic of predicates the alphabet is constituted by the following elements: predicates, variables, functions, constant, connectives, quantifiers and auxiliary symbols. The predicates represent relationships in the speech domain. For example:

- MAN (JOAO)
- MORE_INTELLIGENT (FRANCISCO, PEDRO)
- MORE_INTELLIGENT (FRANCISCO, FATHER(PEDRO)

The set of constants represents the variables.

The functions describe elements and identify them as the only result of the application of a transformation among other elements of the domain [for example: father (PEDRO), mother (father (MICHEL)), and teacher (x)].

Last, the constants represent the elements of the domain of the speech [for example: FRANCISCO, PEDRO, CAROLINA, etc.].

With all the described elements could be capable to build atomic susceptible formulas of an appropriate representation.

The syntax of the language of the logic of predicates is specified by means of a set of primitive symbols and a series of formation rules that allow generating all the formulas belonging to the language.

A signature $\sum = F_\sum \cup P_\sum$ is a set of function symbols f, g, h,... and of symbols of predicate P, Q, R,... each one of them indicating the total number of possible arguments on those that can be applied. We will use the notation: symbol name/total number of possible arguments; for example, $F_\sum = \{c/0, f/1, g/2, h/2\}$ and $P_\sum = \{P/1, Q/2\}$ for the signature $\sum = \{c, f, g, h, P, Q\}$.

The following logical symbols:

- the set of the propositional connectives $\{\neg, \wedge, \vee, \rightarrow, \leftrightarrow\}$

$\wedge$:   With that, for that a well defined formula be right, all and each one of the related components must be right.

$\vee$:   With that, at least one of the related components must be right for that the well defined formula corresponding must be right.

$\neg$:   connective that changes the logical state of an expression.

$\rightarrow$:   that it establishes implication relationships among expressions

$\leftrightarrow$:   connective that indicates the logical equivalence among two well defined formulas

- the universal quantifier $\forall$
- the existential quantifier $\exists$
- the symbol of equality $=$

An infinite set $V = \{x, y, z, u, v, \ldots\}$ of variables

A set of auxiliary symbols formed by open "(" and closed ")" parenthesis and the comma (,).

With all the previous symbols we build the defined alphabet of primitive symbols as $A_\sum = \sum \cup \{\neg, \wedge, \vee, \rightarrow, \leftrightarrow, \forall, \exists, =\} \cup \{(,),\} \cup V$.

The formal language associated to the logic of predicates is the set of all well defined formulas that they can build from the alphabet.

The quantifiers are very important elements of the alphabet that arise as procedure to solve one of the problems mentioned in the logic of propositions. For example:

- Universal quantifier $\forall x$:

It establishes that the well defined formula is certain for all the values that they can take "x".

$$\forall x[PERSON(X) \rightarrow NEEDS\_WATER(X)]$$

- Existential quantifier $\exists x$:

It establishes that exists at least a "x" that makes true to the well defined formula.

$$\exists x[PROPRIETOR(x, CAR) \wedge PROPRIETOR(x, SHIP)]$$

The formal language associated to the logic of predicates is the set of all well defined formulas that they can build from the alphabet.

It can be defined inductively a well defined formula in the following way:

any atomic formula is a well defined formula.

If F and G are well defined formulas, then also they are well defined formulas the following ones:

$$\neg F \quad F \wedge G \quad F \vee G \quad F \rightarrow G \quad F \leftrightarrow G$$

If "x" is a variable, and F is a well defined formula, then also they are well defined formulas the following ones:

$$(\forall x)F \quad (\exists x)F$$

The set of well defined formulas that we are capable to build on a concrete domain constitutes the formal language associated. Thus, and considering the previous definition of well defined formulas, the expression [2]:

$$(\exists x)[(\forall x)[P(x,y) \wedge Q(x,y) \wedge R(x,x) \rightarrow R(x,y)]]$$

is a well defined formula in some domain no certain.

In the logic of predicates can be built two types of expressions:

- Terms that designate individuals of the speech universe
- Formulas that represent enunciated

Therefore, in the logic of predicates are defined formation rules so much for the terms with for the formulas.

Relatively on the semantics, the objective is to attribute meaning to a formula, determining its logical value (true or false) in function of the values of the symbols propositional that it form but in logic of predicates we have to determine the value of terms that they represent the individuals of the speech universe besides and for that we have to use structures and interpretations.

4   Conclusions and Future Work

The propositional languages are insufficient to establish the correction of certain arguments formally due to their inability to reflect valid whose validity cannot demonstrate herself by means of the logical propositional, The problem resides in that this logic type doesn't allow to analyze the internal structure of those enunciated atomic. The language of predicates allows to go farther; their expression capacity is enough for most of the computer science's applications and mathematics. For this reason, we have preferred the logic of predicates as scheme of formal representation of the knowledge. For further work, we will study in-depth the structured methods of representation of the knowledge, that allow, on a part group properties, and on the other hand, to obtain unique descriptions of complex objects, that they are diagrams no formal of representation of the knowledge.

References

1. Baral, C., Delgrande, J., Wotter, F.: Principles of knowledge Representation and Reasoning. kr.org. KR Inc. (2017). Accessed 22 November
2. Bonillo, V.M., Betanzos, A.A., Canosa, M.C., Berdiñas, B.G., Rey, E.M.: Fundamentos de Inteligencia Artificial, Capítulo 4, páginas 87–93, 97 e 98, 111–115, Universidad de la Coruña-España (2000)
3. Gonzales, D.: Verification and Validation. In: The Engineering of Knowledge-Based Systems: Theory and Practice. Prentice-Hall International (1993)
4. Hayes-Roth, F., Waterman, D., Lenat, D.: Building Expert Systems. Addison-Wesley (1983). ISBN0-201-10686-8
5. Ivan, B.: Prolog programming for artificial intelligence, Capítulo 8, páginas 179–181. Addilson-wesley, publishing company (2011)
6. Rich, K.: Artificial Intelligence. McGraw-Hill, New York (1994)
7. Waterman, D.A., Hayes-Roth, F.: Pattern-Directed Inference Systems. Academic Press, New York (1978)

A Review of SEIR-D Agent-Based Model

Farrah Kristel Batista(⊠), Angel Martín del Rey, and Araceli Queiruga-Dios

Institute of Fundamental Physics and Mathematics,
Department of Applied Mathematics, University of Salamanca, Salamanca, Spain
{farrah.batista,delrey,queirugadios}@usal.es

Abstract. Malware is a potential vulnerability for the Internet of Things; it is for this reason that spread of malware on Wireless Sensor Networks has been studied from different perspectives. However, the individual characteristics have not been considered in most of the proposed models. Consequently, Agent-Based Models can be used, as a mathematical tool, to analyse malware propagation. In this work, an ABM is created from three main elements: agents, environment and rules. This article presents a review of an agent-based model to simulate malware spreading in wireless sensor networks.

Keywords: Wireless sensor networks · Agent-based model · Malware · Mathematical models · Cybersecurity

1 Introduction

An Agent-Based Model (ABM) involves the simulation of different types of individuals in a simulated world, with the aim to study the behaviour of these agents based on established rules [21]. These models have three main characteristics: agents, environments and rules.

The individuals called agents are entities with individuals characteristics, capable to execute actions to achieve an objective. They can be a person, animal, plant or a computer. The environment represents a virtual world where agent-agent or agent-environment interactions take place. The rules define a series of behaviours that agents can perform when interacting with another agent or the environment to achieve their goal.

Social environments have been the first application of ABM, to understand the interaction and behaviour of people in an event [9]. Subsequently, ABM has been applied in other areas such as health [3], archeology [24], economics [6] and ecological systems [2]. Consequently, we have decided to implement ABM to cybersecurity, through the study of malware propagation.

The study of malware propagation is the analysis of the behaviour of malware in a network of devices with specific characteristics. Malware uses vulnerabilities in devices, applications, and operating systems to achieve its goal: distribute malicious code, information theft, denial of service (DoS), or fraud.

© Springer Nature Switzerland AG 2020
E. Herrera-Viedma et al. (Eds.): DCAI 2019, AISC 1004, pp. 133–140, 2020.
https://doi.org/10.1007/978-3-030-23946-6_15

The best-known types of malware are computer viruses, computer worms, trojans, rootkits, botnets, and spyware [17]. The spread of malware has been studied in different types of networks, e.g. computer networks, mobile networks, wireless networks or wireless sensor networks.

Wireless Sensor Networks (WSN) are a set of sensor devices deployed in a given area that form a network without a pre-established architecture. The usefulness of these networks comes from the collaboration of a large number of devices, called nodes, to generate data and information that is sent to a base station [7].

This article presents a review of a novel agent-based model for simulate malware propagation in wireless sensor networks.

The article is structured in the following way: in Sect. 2 the related work is introduced. In Sect. 3 the review of the SEIR-D model is presented. Finally, in Sect. 4 the conclusions are shown.

2   Related Works

A bibliographic search has been made of the works that have proposed mathematical models to study the propagation of malware in WSN. These papers have been published between January 2014 and January 2019. A brief classification of the mathematical models proposed in these works is shown in Table 1. This classification is based on the type of compartment, the type of model and the mathematical tool used in the model.

The model proposed in [11] is the basis of mathematical epidemiology, and almost all subsequent models are based on it. The models can be classified as global or individual, discrete or continuous, deterministic or stochastic. In addition, the mathematical tools have been defined as a partial differential equation (PDE), ordinary differential equations (ODE), Markov chains or cellular automata (CA).

In summary, individual and global models with different mathematical tools have been used for the study of malware propagation.

Some advantages of ABMs are that they allow simulating learning at both individual and population levels, provide high scalability, heterogeneity of agents and environments, model behaviours can be adjusted through rules, and the programming environment is easy to understand [9].

3   Agent-Based Model Proposed

WSNs are a set of devices that work together to do a task; each device provides the resources that they have available. These devices or sensors are characterised by have not advanced security measures to protect the data and information that they handled. Also, the wireless communication channel adds another risk factor, because it can be intercepted more easily than a wired channel.

Zigbee is the most commonly used protocol in WSN; this protocol is also used in IoT devices. Researches have shown that this protocol can be exposed to

Table 1. Classification of the mathematical models.

Work	Compartment	Type of model			Tool
[25]	SEIR-V	Global	Continuous	Deterministic	ODE
[5]	SIR	Individual	Discrete	Stochastic	CAs
[14]	SI	Global	Discrete	Stochastic	Markov Chain
[12]	SIRC, SIQRC, SIRVC	Global	Continuous	Deterministic	ODEs
[13]	SEIR	Global	Continuous	Deterministic	ODEs
[8]	SIRS	Global	Continuous	Deterministic	ODEs
[10]	SIR	Individual	Discrete	Deterministic	CAs
[26]	SIRS	Global	Continuous	Deterministic	ODEs
[27]	SIR	Global	Continuous	Deterministic	PDEs
[19]	SIR	Global	Discrete	Stochastic	Markov Chain
[18]	SIR	Individual	Discrete	Deterministic	CAs
[20]	SIS	Global	Discrete	Stochastic	Markov Chain
[22]	SI	Individual	Discrete	Stochastic	Markov Chain
[23]	SIR	Individual	Discrete	Deterministic	CAs

attacks such as eavesdropping, Denial of Service (DoS), node compromise, sinkhole and wormhole attacks, physical attack [4], and matched protocol attack [15].

The improvement of security to WSN or IoT is an issue of concern to the scientific community, due to the considerable growth in demand and use of these devices. For this reason, the study of the propagation of malware is an issue that allows improving the techniques and security measures that can be implemented in a WSN.

In this section, we will describe the steps we have taken into account to develop an agent-based model for malware propagation. We have divided this section as follows: in Subsect. 3.1 the agents involved in the model are mentioned, in Subsect. 3.2 the coefficients associated to the sensor agents are detailed, and finally in Subsect. 3.3 the transition rules that allow the sensor agents to change their states are described.

3.1   Agents

We have started the analysis by identifying the main agents that may be involved in the propagation process. These agents will be described below:

- Sensors: they are the core element of the WSN. These agents collect the data directly from the environment, process the information, and some of them are able to transmit the processed information to the base station. The sensors can be classified into sensor nodes, router/cluster-head nodes and sink nodes. Each sensor has unique characteristics according to the function it performs within the network.

- Malware: it is malicious code designed to harm sensors and spread across the network. The target of the malware can be information exfiltration, denial-of-service (DoS) or fraud.
- Network topology: defines how sensors are organised to communicate with each other and with the base station. The topology can be mesh, star or hybrid. Also, the implemented protocols include Self-Organizing Protocol (SOP) and Energy Clustering and Routing (EECR).
- The phenomenon of interest: represents the physical environment. It is the environment monitored by sensors to gather information. The phenomenon can be military, industrial, public health, environmental, daily activities or multimedia.
- Human action: refers to the interaction of users, technicians, administrators or attackers with sensors or the network.
- Devices: are those devices that can have contact with the sensor network. It can be classified both external devices (e.g. external hard drive, memory stick, USB), and computational devices such as base station, servers or computers.

3.2 Coefficients

The next step consists of defining the coefficients that determine the parameters. The coefficients that have been defined are the following:

- Infection coefficient: establishes the probability that a sensor agent has the computational capacity to be attacked and infected by a malware designed for WSN. This coefficient depends on the computational characteristics of the sensor agent, prioritising the security measures that have been implemented in the node or network.
- Transmission coefficient: it indicates whether a sensor agent has the computational and communication capacity to transmit the malware that has infected it. This coefficient depends on the characteristics of the sensor, especially in the data transmission channel or antenna, and the topology that determines the number of neighbours.
- Detection coefficient: determines if a sensor agent has the capacity to detect that it has been infected by malware, send a notification, and begin the process of remove malware from the sensor agent. This coefficient depends on the computational characteristics of the sensor agent and the maintenance that has been applied by technicians and administrators.
- Recovery coefficient: indicates whether a sensor agent has the capacity to resume its normal operating state after the malware has been removed. This coefficient depends on the computational characteristics of the sensor agent and the network topology.
- Maintenance coefficient: establishes if the maintenance has been applied to a sensor agent by technicians or administrators. Maintenance may consist of recharging the battery, software upgrades, or replace a physical part of the sensor.

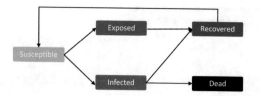

Fig. 1. Scheme of SEIR-D model.

– Energy coefficient: determines if a sensor agent has the minimum energy required to continue normal operation, or whether the sensor has stopped work. This coefficient depends on the battery or power source of the sensor.
– Malware coefficient: indicates whether malware has been designed to attack the sensors of a WSN. This coefficient depends on the characteristics of the malware, such as type, target and propagation mechanism.

3.3 Rules

Transition rules define the behaviour of agents in different situations, the interaction with other agents and the environment. These rules define when and why the agents must change from one state to another. The mathematical epidemiology model has been necessary to define the transition rules. In this case, the following states have been defined (see Fig. 1).

– Susceptible: a sensor agent in a susceptible state has not been infected by malware, but it has the computational characteristics to be infected.
– Exposed: a sensor agent can become exposed when malware has been hosted, but it is not able to transmit the malware to its neighbours.
– Infected: a sensor agent with an infected state has been successfully infected by malware. Malware exploits its activity, and it can spread.
– Recovered: a sensor agent in the recovered state has been able to remove the malware and return to its normal operating state.
– Dead: a sensor agent is dead when its power has been quickly depleted due it has been infected by malware.

Then, the following transition rules have been defined for the sensor agents:

– Susceptible to Infected: an agent in susceptible state changes to infected state when malware has exploited some vulnerability and successfully infected the sensor agent. The coefficients related are the infection, energy and malware.
– Susceptible to Exposed: a sensor agent goes from susceptible state to exposed state when it has been infected by malware but does not have the computational capacity to transmit the malware to neighbouring sensor agents. The coefficients related are the transmission, energy and malware.
– Infected to Recovered: a sensor agent can go from infected state to recovered state when the malware infection has been removed and returned to its normal operative state. The coefficients related are the detection, recovery and maintenance.

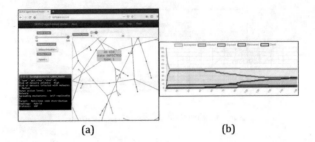

(a)   (b)

Fig. 2. Illustrative example of (a) Mesa Framework in console and browser and (b) graphic of compartment evolution.

– Exposed to Recovered: a sensor agent goes from exposed to recovered state when the malware infection has been removed and returned to its normal operating state. The coefficients related are the recovery, maintenance and energy.
– Recovered to Susceptible: a sensor agent changes from recovered to susceptible state when all sensor agents on the same network have successfully removed the malware infection and returned to the normal operative state. However, the infection cycle can be restart with another attack or sample malware. The coefficients related are the maintenance and malware.
– Dead: a sensor agent becomes dead when its battery or power supply has stopped working, and it has shut down. The coefficient related is energy.

3.4 Simulation

In order to simulate this model, it is necessary to use specialised MBA simulation software. First, we will analyse the current software to determine which fit the needs of our model.

In [1], the authors present an exhaustive review of these types of software, by area of application and easy of use and scalability of the model. After evaluating different options, we have selected the Framework Mesa [16], developed in Python; this programming language allows the adjustment of the framework to the needs of the user. This framework has been selected based on the number of nodes supported; moreover, the simulation is done in real time and does not require a large number of IT resources. An illustrative example of the results can be visualised through the console or a browser (see Fig. 2).

The installation of the framework has been used Ubuntu Linux 16.4 distribution in a Virtual Machine, with an Intel i5 processor, 2 GB RAM, 10 GB HDD and Internet connection. The installation of Mesa has been done through an installation command in the Linux console, following the user manual.

4 Conclusions

The analysis of malware propagation in WSN through ABM require to be completed a series of steps. First of all, we must define the agents and the environment, then the coefficients and rules, create the simulations, and finally, validate and verify the results. In this work, we have shown the steps that have been performed and the characteristics that have been defined for the creation of this model.

ABMs can provide a more comprehensive and precise view of malware behaviour at WSN when each of the steps is defined correctly. Malware, sensors and the environment can be adjusted with characteristics of a real environment to obtain more precise results.

As future work, we propose the analysis of the results that can be obtained with the simulation of this model in three different environments, e.g. military and industrial, health and environmental, finally daily activities and multimedia.

Acknowledgements. This research has been partially supported by Ministerio de Ciencia, Innovación y Universidades (MCIU, Spain), Agenda Estatal de Investigación (AEI, Spain), and Fondo Europeo de Desarrollo Regional (FEDER, UE) under project with reference TIN2017-84844-C2-2-R (MAGERAN) and the project with reference SA054G18 supported by Consejería de Educación (Junta de Castilla y León, Spain). F.K. Batista has supported by IFARHU-SENACYT scholarship program (Panama).

References

1. Abar, S., Theodoropoulos, G.K., Lemarinier, P., O'Hare, G.M.: Agent based modelling and simulation tools: a review of the state-of-art software. Comput. Sci. Rev. **24**, 13–33 (2017)
2. Anderson, T.M., Dragićević, S.: Network-agent based model for simulating the dynamic spatial network structure of complex ecological systems. Ecol. Modell. **389**, 19–32 (2018)
3. Arifin, S.N., Madey, G.R., Collins, F.H.: Spatial Agent-Based Simulation Modeling in Public Health: Design, Implementation, and Applications for Malaria Epidemiology. Wiley, New York (2016)
4. Bin Karnain, A., Bin Zakaria, Z.: A review on ZigBee security enhancement in smart home environment. In: 2nd International Conference on Information Science and Security (ICISS), pp. 1–4. IEEE (2015)
5. Chizari, H., Zulkurnain, A.U.: Modelling malware response in wireless sensor networks using stochastic cellular automata. J. Mobile Embed. Distrib. Syst. **6**(4), 159–166 (2014)
6. Chu, Z., Yang, B., Ha, C.Y., Ahn, K.: Modeling GDP fluctuations with agent-based model. Physica A **503**, 572–581 (2018)
7. Conti, M.: Secure Wireless Sensor Networks: Threats and Solutions, vol. 65. Springer, New York (2015)
8. Feng, L., Song, L., Zhao, Q., Wang, H.: Modeling and stability analysis of worm propagation in wireless sensor network. Math. Probl. Eng. (2015)

9. Helbing, D.: Social Self-Organization: Agent-Based Simulations and Experiments to Study Emergent Social Behavior. Springer, Heidelberg (2012)
10. Hu, J., Song, Y.: The model of malware propagation in wireless sensor networks with regional detection mechanism. In: China Conference on Wireless Sensor Networks, pp. 651–662. Springer (2014)
11. Kermack, W.O., Mckendrick, A.G.: A contribution to the mathematical theory of epidemics. Proc. R. Soc. Lond. **115**, 700–721 (1927)
12. Keshri, N., Mishra, B.K.: Optimal control model for attack of worms in wireless sensor network. Int. J. Grid Distrib. Comput. **7**, 251–272 (2014)
13. Keshri, N., Mishra, B.K.: Two time-delay dynamic model on the transmission of malicious signals in wireless sensor network. Chaos Solitons Fractals **68**, 151–158 (2014)
14. Li, Q., Zhang, B., Cui, L., Fan, Z., Athanasios, V.V.: Epidemics on small worlds of tree-based wireless sensor networks. J. Syst. Sci. Complex **27**(6), 1095–1120 (2014)
15. O'Mahony, G.D., Harris, P.J., Murphy, C.C.: Analyzing the vulnerability of wireless sensor networks to a malicious matched protocol attack. In: 2018 International Carnahan Conference on Security Technology (ICCST), pp. 1–5. IEEE (2018)
16. Project Mesa Team: Mesa: Agent-Based Modeling in Python 3+ (2018). https://github.com/projectmesa/mesa/
17. Razak, M.F.A., Anuar, N.B., Salleh, R., Firdaus, A.: The rise of "malware": bibliometric analysis of malware study. J. Netw. Comput. Appl. **75**, 58–76 (2016)
18. del Rey, A.M., Guillén, J.H., Sánchez, G.R.: Modeling malware propagation in wireless sensor networks with individual-based models. In: Conference of the Spanish Association for Artificial Intelligence, pp. 194–203. Springer (2016)
19. Shen, S., Huang, L., Liu, J., Champion, A.C., Yu, S., Cao, Q.: Reliability evaluation for clustered WSNs under malware propagation. Sensors **16**(6), 855 (2016)
20. Shen, S., Ma, H., Fan, E., Hu, K., Yu, S., Liu, J., Cao, Q.: A non-cooperative non-zero-sum game-based dependability assessment of heterogeneous wsns with malware diffusion. J. Netw. Comput. Appl. **91**, 26–35 (2017)
21. Siegfried, R.: Modeling and Simulation of Complex Systems: A Framework for Efficient Agent-Based Modeling and Simulation. Springer, Berlin (2014)
22. Wang, T., Wu, Q., Wen, S., Cai, Y., Tian, H., Chen, Y., Wang, B.: Propagation modeling and defending of a mobile sensor worm in wireless sensor and actuator networks. Sensors **17**(1), 139 (2017)
23. Wang, Y., Li, D., Dong, N.: Cellular automata malware propagation model for WSN based on multi-player evolutionary game. IET Netw. **7**(3), 129–135 (2017)
24. Wurzer, G., Kowarik, K., Reschreiter, H.: Agent-based Modeling and Aimulation in Archaeology. Springer, Cham (2015)
25. Zhang, Z., Si, F.: Dynamics of a delayed SEIRS-V model on the transmission of worms in a wireless sensor network. Adv. Differ. Equations **2014**(1), 295 (2014)
26. Zhu, L., Zhao, H.: Dynamical analysis and optimal control for a malware propagation model in an information network. Neurocomputing **149**, 1370–1386 (2015)
27. Zhu, L., Zhao, H., Wang, X.: Stability and bifurcation analysis in a delayed reaction-diffusion malware propagation model. Comput. Math. Appl. **69**(8), 852–875 (2015)

Proposed Models for Advanced Persistent Threat Detection: A Review

Santiago Quintero-Bonilla$^{(\boxtimes)}$ and Angel Martín del Rey

Institute of Fundamental Physics and Mathematics,
Department of Applied Mathematics, University of Salamanca, Salamanca, Spain
{santiago.quintero,delrey}@usal.es

Abstract. Advanced Persistent Threat is a sophisticated, targeted attack. This threat represents a risk to all organisations, specifically if they manage sensitive data or critical infrastructures. Recently, the analysis of these threats has caught the attention of the scientific community. Researchers have studied the behaviour of this threat to create models and tools that allow early detection of these attacks. The use of Artificial Intelligence can help to detect, alert and automatically predict these types of threats and reduce the time the attacker can stay on a network organisation. The objective of this work is a review of the proposed models to identify the tools and methods that they have used.

Keywords: Artificial intelligence · Machine learning ·
Advanced persistent threats · Malware · Detection · Cybersecurity

1 Introduction

Advanced Persistent Threats (APT) has been defined in several ways. In this work, we have been adopted the definition given by the US National Institute of Standards and Technology (NIST): "An adversary that possesses sophisticated levels of expertise and significant resources which allow it to create opportunities to achieve its objectives by using multiple attack vectors (e.g., cyber, physical, and deception)" [15].

Attackers study the behaviour of their victim to use them as the initial intrusion. They will look for ways to move laterally within the network, in search of users with high privileges, and access services with sensitive information [14].

The attack vectors can differ considering the particular and specific characteristics of the victim profiles. In this sense, Social Engineering plays an important role [11]. Attackers perform an intelligence gathering phase, and this generates attacks on mobile devices, exploit vulnerabilities, and even use the distribution of infected USB, e.g. the Stuxnet case [5], Duqu 2.0 [10] and Industroyer [3].

The characteristics of APT are the following: they have specific targets, sophisticated and highly organised attacks, endowed with a large number of resources. This type of attack is designed to remain within a computer network undetected for an extended period, usually in order to extract confidential

© Springer Nature Switzerland AG 2020
E. Herrera-Viedma et al. (Eds.): DCAI 2019, AISC 1004, pp. 141–148, 2020.
https://doi.org/10.1007/978-3-030-23946-6_16

information [17]. These features have made detection inefficient using traditional detection methods. Several companies have submitted reports [2,4,13], demonstrating the techniques used by APTs against government institutions, companies, industry sectors and individuals.

Each APT campaign is customised for each victim or organisation, and it has a unique life cycle. The report of the company FireEye about APT1 [12], "Anatomy of Persistent Advanced Threats", proposes a life cycle consisting of the following eigth phases: (1) Initial Recon, (2) Initial Compromise, (3) Establish Foothold (4) Escalate Privileges, (5) Internal Recon, (6) Move Laterally, (7) Maintain Presence (8) Complete Mission; In this life cycle, from the phase 4 to phase 7, the attackers maintain an extended persistence access in the attacked environment. In addition, the authors in [21] has analyzed 22 APT campaigns and they have identified three main phases of the life cycle of an APT: (1) initial compromise, (2) lateral movement and (3) command and control. This work allows visualising the relevant attributes with the techniques and methods used in an APT attack to propose prevention and detection approaches.

This article presents a review of the frameworks and models proposed for detection of APT attack.

The rest of paper is organised as follows: in Sect. 2 the Machine Learning methods are detailed, in Sect. 3 the related work is introduced and finally in Sect. 4 the conclusion is presented.

2   Machine Learning Methods

This section describes Machine Learning methods and the algorithms used for the different APT detection approaches that will help to understand the later sections.

Machine Learning (ML) is a tool that has begun to show usefulness in many scientific areas for problem-solving, due to the significant advantages of adaptability and scalability to the unknown [9]. Some examples of machine learning applications in cybersecurity are phishing detection, network intrusion detection, checking protocol security properties, cryptography and spam detection in social networks [6]. ML uses two types of techniques: supervised learning (training the model with both known input and output data to predict future results), and unsupervised learning (extracting the description of hidden structures from unlabelled data) [22]. The ML algorithms have been used for APT attack detection, and prediction approaches are the following:

- Desicion tree (DT) is a predictive model, it divides a complex desicion process into groups of less complicated decision [17]. Commonly used DT models are classification and regression trees (CART), ID3 and C4.5 [22].
- Support vector machine (SVM) analyses data and identifies patterns. It is a classifier based on finding a separating hyperplane in the feature space between two classes in such a way that the distance between the hyperplane and the closest data points of each class is maximised [1].

- k-nearest neighbours (k-NN) classifier is based on a distances function that measures the differences or similarity between two instances. The variation of k in k-NN represents the number of neighbour instances that need be compared, the results of the votes of the majority class element is considered the class [19,22].
- Ensemble learning combines multiple hypotheses, hoping to form a better prediction alone [1].
- Genetic programming (GP) is a subclass of genetic algorithms, which use replication and mutation to develop structures [16].
- Dynamic bayesian game model (DBG-model) is based on the diverse requirements of the application, and the game models are typically classified on two viewpoints: (1) dynamic and static game and (2) complete and incomplete information game [16].
- Correlation fractal dimension (FD) algorithm requires a reference dataset of the labelled features. Each new data point is classified as anomalous or benign by comparing the correlaction FD of the corresponding dataset [19].

3   Background

The detection of APT has been a challenge for current defence systems in cyber-security. Different research approaches are being carried out to discuss this type of attacks. Recently, frameworks or models have been developed to help understand the behaviour of APT attacks. In what follows we present a brief analysis of existing models (see Table 1).

The authors in [7], present a novel machine learning-based system called MLAPT. This model consists of generating early warnings to detect an APT attack. The MLAPT is based on the analysis of a six-phase APT life cycle: (1) Intelligence Gathering, (2) Point of entry, (3) C&C Communication, (4) Lateral Movement, (5) Asset/Data Discovery, (6) Data Exfiltration. An individual detection model is inefficient at detecting APT attacks. For this reason, the authors seek to create a correlation framework between the different detection modules. These modules create alerts, and then they are analysed through ML algorithms. MLAPT works in three phases:

- Threat Detection: in this phase, network traffic is scanned through eight detection modules to detect techniques used by APTs. The detection modules are: Disguised exe File Detection (DeFD), Malicious File Hash Detection (MFHD), Malicious Domain Name Detection (MDND), Malicious IP Address Detection (MIPD), Malicious SSL Certificate Detection (MSSLD), Domain Flux Detection (DFD), Scan Detection (SD) and Tor Connection Detection (TorCD). The output of this phase are alerts known as events.
- Alert Correlation: in this phase, the alerts generated by the detection modules are correlated to find an APT attack. This phase is carried out in three steps: (1) alert filter, (2) alert clustering and (3) correlation indexing. At the end of the analysis can be generated two types of alerts: *apt_full_scenario_alert* and *apt_sub_scenario_alert*.

Table 1. Characteristics of frameworks proposed.

Paper	Framework	Design of the framework	APT life cycle
[7]	MLAPT	(1) Threat detection phase	6 phases
		(2) Alert correlation phase	
		(3) Attack prediction phase	
[16]	DFA-AD	(1) Initial phase: network traffic Intermediate phase: Classificator	Non-specified
		(2) Second phase: event correlation module	
		(3) Third phase: Voting services	
[19]	Fractal based anomaly	Anomaly classification algorithms:	Non-specified
		(1) k-NN	
		(2) Correlation fractal dimension based	
[20]	Detection based on dynamic analysis	Components:	3 phases
		(1) Network traffic redirection module	
		(2) User agent	
		(3) Reconstruction module	
		(4) Dynamic analysis module	
		(5) Decision module	
[18]	Detection based on big data	k-NN algorithm based on Mahout Phases:	Non-specified
		(1) Retrieve	
		(2) Reuse	
		(3) Revise	
		(4) Retain	
[8]	Context-based detection	Components:	6 phases
		(1) Pyramid attack levels	
		(2) Planes and events	
		(3) Correlation rules	
		(4) Detection rules	

– Attack prediction: this phase uses a machine learning-based prediction module. This process is completed in three steps (1) dataset preparation (2) training the prediction model and (3) using the model for prediction. In this module, the ML algorithms have been used are decision tree learning, support vector machine, k-nearest neighbours and ensemble learning.
  As a result of this work, the authors indicate that the average of true positives is 81.8% and an average of 4.5% of false positives.

The authors in [16] propose a novel distributed framework architecture for APT detection called DFA-AD. This framework is based on multiple parallel classifiers, which classify events in a distributed environment and the correlation of events between them. Each of these classifiers focuses on detecting the techniques used by the APT to carry out the attack.

Intrusion detection is realised in a distributed environment on the trusted platform module (TPM). A TPM can be integrated into the motherboard of a computer to implement secure communication within a network. DFA-AD is designed in three phases:

- Network traffic: in this phase, network traffic packets are collected, processed and analysed to identify all possible strategies that can be used in the life cycle of an APT attack. Also, the authors use four different classifiers as a method of recognition: genetic programming, classification and regression trees, support vector machines and dynamic bayesian game model. The output of these classification methods is transferred to the event correlation phase.
- Correlation event: in this phase, all the events generated by the recognition classifiers used in the previous phase are gathered to be evaluated using specific rules by an administrator who will warn about the detection of an APT attack.
- Voting service: after all the information generated in the previous modules, a voting service will analyse and determine if the system should generate an alert about an APT attack. The authors suggest that the module contributes to a decrease in the rate of false positives and improves the accuracy of the detection system.

This framework has had a good detection rate of 98.5% with 0.024 false positives.

In the article [19], fractal-based anomaly classification mechanisms are presented to reduce false positives and false negatives.

The first step was to combine two datasets — one dataset with normal traffic packets and the another with APT attack network traffic packets. Then, an analysis of the TCP session data has been made to determine the characteristics of the vector. The classification of the compromise is specified in two metrics: (1) the total number of packets transferred during a single TCP session; (2) the duration of the complete session. Consequently, the dataset noise has been eliminated in two steps: (1) eliminate the packets that have zero length; (2) remove retransmitted packets. At the end of this pre-processing phase, the data is ready to be used in the algorithms.

Two anomaly classification algorithms have been used:

- k-nearest neighbours (k-NN) is a supervised learning algorithm, with a class of instant based learner.
- Correlation fractal dimension (FD) is based on exploiting the multifractal nature of the Internet data series.

The results of the fractal correlation dimension have improved by 12% accuracy and f-measure concerning the k-NN algorithm. The authors claimed the fractal dimension based algorithm is better than the Euclidean dimension based algorithm. The multifractal analysis extracts hidden information about a measurement that is not possible in a mono scale analysis.

The authors in [8], have designed a detection framework based on an attack pyramid model. This model works in the following way: correlation rules in contexts relate the events collected by each level of the pyramid. These contexts are exported to the alert system. This system applies the detection rules using a signature database. An evaluation of the detection rules is performed to update the levels of confidence and risk for each context. Threshold levels are checked based on confidence and risk results. If the threshold levels are in the alarm zone, an alarm is triggered and the APT incident response is initiated. Besides, the security analyst is notified to initiate the investigation of the event.

The authors in [20], propose a framework for dynamically reconstructing the network data stream to detect APT without affecting the typical workflow. This framework is built on five main components:

- Network traffic redirection module: this module copies the current network traffic and redirects it to the reconstruction module.
- User-agent: this module provides auxiliary information on the host for the reconstruction and decision module.
- Reconstruction module: restores data with malicious content and sends it to the dynamic analysis module.
- Dynamic analysis module: this module use the auxiliary information provided by the user-agent module and the data from the reconstruction module. It is possible to create a virtual environment similar to any host in the network.
- Decision module: this module integrates the above information and obtains a conclusion according to predefined criteria.

The authors in [18], propose an APT detection system, based on the Big Data architecture process. This system consists of four steps:

- Architecture of the APT system: in this phase, information related to the network data and the information in the system registers are collected and analysed.
- Processing technology: in this phase, to improve the analysis of an APT attack, is used the computational performance of the Hadoop[1] cluster.
- APT Analysis Technology: this phase includes the detection of malicious attacks through known vulnerabilities and suspicious connections with anomalous behaviour.
- APT detection algorithm based on Mahout: the authors indicate that the Mahout[2] library is suitable for APT detection because it can process large amounts of data, and the k-NN machine learning algorithm found within the Mahout project is used. This model has four phases: Retrieve, Reuse, Revise, Retain.

The results obtained during the test environment show that the capacity of the system to process the data can be up to 10 TB, together with the performance analysis of the data to classify an APT can be 2000 samples in parallel, to satisfy the requirements of APT detection.

---

[1] Apache Hadoop™ - https://hadoop.apache.org/.
[2] Apache Mahout™ - https://mahout.apache.org/.

4   Conclusion

In this article, we have reviewed the Machine Learning processes, techniques and algorithms that have been used in different approaches for the detection and prediction of APT attacks. We have been identified that the ML algorithms used in these approaches are supervised learning algorithms. This type of supervised learning allows data labelled from input data. Commonly algorithms used in these approaches to detect APT attacks are k-NN, SVM and Decision tree. The k-NN algorithm has been used in four of these approaches and has obtained the best results for APT attack prediction with a high rate of true positives.

In future work, we will propose a novel framework that implements algorithms to help process data generated and collected by security detection tools. Furthermore, the implementation of Machine Learning techniques in the event analysis would be advantageous, because a well-trained algorithm can learn to generate alerts to predict or not when an APT attack is executed. Also, the prediction step can be detected attack patterns, and anomalous connections with the data processed to make decisions. These decisions can be created rules to block traffic from a suspicious source and generate reports to the network administrators.

Acknowledgements. This research has been partially supported by Ministerio de Ciencia, Innovación y Universidades (MCIU, Spain), Agenda Estatal de Investigación (AEI, Spain), and Fondo Europeo de Desarrollo Regional (FEDER, UE) under project with reference TIN2017-84844-C2-2-R (MAGERAN) and the project with reference SA054G18 supported by Consejería de Educación (Junta de Castilla y León, Spain).

S. Quintero-Bonilla has been supported by IFARHU-SENACYT scholarship program (Panama).

References

1. Buczak, A.L., Guven, E.: A survey of data mining and machine learning methods for cyber security intrusion detection. IEEE Commun. Surv. Tutorials **18**(2), 1153–1176 (2016)
2. Check Point Research: Global Cyber Attack Trends Report. Technical report (2017)
3. Cherepanov, A.: WIN32/INDUSTROYER: A new threat for industrial control systems. Technical report (2017)
4. Cisco Systems, Inc.: Midyear Cybersecurity Report. Technical report (2017)
5. Falliere, N., Murchu, L.O., Chien, E.: W32. stuxnet dossier. White Pap. Symantec Corp., Secur. Response **5**(6), 29 (2011)
6. Ford, V., Siraj, A.: Applications of machine learning in cyber security. In: Proceedings of the 27th International Conference on Computer Applications in Industry and Engineering (2014)
7. Ghafir, I., Hammoudeh, M., Prenosil, V., Han, L., Hegarty, R., Rabie, K., Aparicio-Navarro, F.J.: Detection of advanced persistent threat using machine-learning correlation analysis. Futur. Gener. Comput. Syst. **89**, 349–359 (2018)

8. Giura, P., Wang, W.: A context-based detection framework for advanced persistent threats. In: Proceedings of the 2012 ASE International Conference on Cyber Security, CyberSecurity 2012, SocialInformatics, pp. 69–74 (2012)
9. Jang-Jaccard, J., Nepal, S.: A survey of emerging threats in cybersecurity. J. Comput. Syst. Sci. **80**(5), 973–993 (2014)
10. Kasperky Lab: The Duqu 2.0 - Technical Details (V2.1). Technical report June (2015)
11. Krombholz, K., Hobel, H., Huber, M., Weippl, E.: Advanced social engineering attacks. J. Inf. Secur. Appl. **22**, 113–122 (2015)
12. Mandiant: APT1 Exposing One of China's Cyber Espionage Units. Technical report (2013)
13. Mandiant: M-Trends 2017: A view from the front lines. Technical report (2017)
14. Navarro, J., Deruyver, A., Parrend, P.: A systematic survey on multi-step attack detection. Comput. Secur. **76**, 214–249 (2018)
15. NIST: Managing information security risk: Organization, mission, and information system view. Special Publication 800-839 (2011)
16. Sharma, P.K., Moon, S.Y., Moon, D., Park, J.H.: DFA-AD: a distributed framework architecture for the detection of advanced persistent threats. Cluster Comput. **20**(1), 597–609 (2017)
17. Shenwen, L., Yingbo, L., Xiongjie, D.: Study and research of apt detection technology based on big data processing architecture. In: 5th International Conference on Electronics Information and Emergency Communication (ICEIEC), pp. 313–316. IEEE (2015)
18. Shenwen, L., Yingbo, L., Xiongjie, D.: Study and research of apt detection technology based on big data processing architecture. In: 2015 5th International Conference on Electronics Information and Emergency Communication (ICEIEC), pp. 313–316. IEEE (2015)
19. Siddiqui, S., Khan, M.S., Ferens, K., Kinsner, W.: Detecting advanced persistent threats using fractal dimension based machine learning classification. In: Proceedings of the 2016 ACM on International Workshop on Security And Privacy Analytics - IWSPA 2016, pp. 64–69. ACM Press (2016)
20. Su, Y., Li, M., Tang, C., Shen, R.: A framework of apt detection based on dynamic analysis. In: 2015 4th National Conference on Electrical, Electronics and Computer Engineering. Atlantis Press (2015)
21. Ussath, M., Jaeger, D., Cheng, F., Meinel, C.: Advanced persistent threats: behind the scenes. In: Annual Conference on Information Science and Systems (CISS), pp. 181–186. IEEE (2016)
22. Xin, Y., Kong, L., Liu, Z., Chen, Y., Li, Y., Zhu, H., Gao, M., Hou, H., Wang, C.: Machine learning and deep learning methods for cybersecurity. IEEE Access **6**, 35365–35381 (2018)

A Global Classifier Implementation for Detecting Anomalies by Using One-Class Techniques over a Laboratory Plant

Esteban Jove[1,2(✉)], José-Luis Casteleiro-Roca[1], Héctor Quintián[1],
Juan-Albino Méndez-Pérez[2], and José Luis Calvo-Rolle[1]

[1] Department of Industrial Engineering, University of A Coruña,
Avda. 19 de febrero s/n, 15405 Ferrol, A Coruña, Spain
esteban.jove@udc.com
[2] Department of Computer Science and System Engineering,
University of La Laguna, Avda. Astrof. Francisco Sánchez s/n,
38200 S/C de Tenerife, Spain

Abstract. The energy and the product optimization of the industrial processes has played a key role during last decades. In this field, the appearance of any kind of anomaly may represent an important issue. Then, anomaly detection in an industrial plant is specially relevant.

In this work, the anomaly detection over level plant control is achieved, by using three one class intelligent techniques. Different global classifiers are trained and tested with real data from a laboratory plant, whose main aim is to control the tank liquid level. The results of each classifier are assessed and validated with real anomalies, leading to good results, in general terms.

Keywords: Fault detection · One-class · ACH · Autoencoder · SVM

1 Introduction

The significant advances in the industrial field, in terms of digitalization or instrumentation, has lead to the development and the optimization of many industrial processes [13,25]. Also, additional factors, such as the quality and the reliability standards, or the promotion of energy efficiency policies, are also taken into consideration in an industrial process.

In this context, a good system operation plays an important role. Hence, an early detection of anomalies becomes very important, paying special attention to high cost and safety-critical processes [24]. The appearance of an anomaly can be caused by many different events like sensor misreadings, actuator malfunctioning and changes in different process parameters [3,9,20].

In addition to industrial systems, anomaly detection techniques can be applied in many different fields, such as: electric systems, medicine tasks or fraud detection in bank accounts [1,6,11,18,33].

© Springer Nature Switzerland AG 2020
E. Herrera-Viedma et al. (Eds.): DCAI 2019, AISC 1004, pp. 149–160, 2020.
https://doi.org/10.1007/978-3-030-23946-6_17

From a generic point of view, anomalies are defined as data patterns that present an unexpected behavior in an specific application [7]. In an industrial process, a deviation, of at least one variable or feature of the system from its usual and right condition, is considered anomaly.

However, the process of classify a new data as normal or anomalous, must face several problems. The main issue lies in the task of selecting a limit that separates properly normal and anomalous data. This process can be harder, if the dataset contains noise, or if the abnormal data conform a relatively big group, because the risk of considering it normal is significant. Also, in some applications, the systems evolve with their use, changing some of their parameters, although this could not imply an anomaly.

Then, given a specific dataset, the anomaly detection can be faced in three different ways, depending on the previous information about it [7,14].

- Type A: in this type of anomaly detection, the dataset has only normal data or normal data with a few anomalies. Then, the classifier is trained with normal data, and an anomaly is identified when a data with different features arrives.
- Type B: the dataset is not prelabeled, because there is not information about the nature of the initial dataset. Hence, unsupervised algorithms are applied to separate normal and anomalous data.
- Type C: the initial dataset can be divided and prelabeled between normal and abnormal data. Therefore, supervised algorithms are applied to obtain representative models of both dataset.

This works deals the anomaly detection in an industrial plant used to control the fluid level in a tank. In this case, a global classifier is implemented using a dataset from three different operating points. Therefore, the classifiers are trained with normal data, where the anomalies are not statistically representative.

To obtain one-class classifiers capable of detecting anomalies on the laboratory plant, the techniques Support Vector Machine (SVM), Aproximate Convex Hull (ACH) and Autoencoder are used.

The proposed method was assessed and validated on a laboratory real plant used to control the liquid level in a tank. To validate the method, real anomalies were generated by opening the output valve.

The rest of the document is structured as follows. Section 2 describes briefly the case of study. After Sect. 2, the techniques applied to achieve the fault detection one-class classification are explained. Section 4 details the experiments and achieved results and finally, the conclusions and future works are exposed in Sect. 5.

2   Case of Study

This section focuses on the description of the plant under study. The general features of the dataset are described below.

2.1   Control Level Plant

The scheme of the laboratory plant under study is shown in Fig. 1. This plant is designed with industrial equipment, and whose objective is to control the liquid level of a tank measured by an ultrasonic sensor. The fluid is initially in a lower tank, and it is pumped using a three-phase pump driven by a variable frequency driver. The fluid flow depends on the pump speed. The objective tank, has two built in output valves, one of them is a manual and the other one is electric. They are used as a pathway to turn the fluid back to the initial tank.

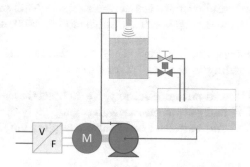

Fig. 1. Scheme of the control level plant

2.2   Control System Implementation

The virtual controller Fig. 2 was implemented using Matlab software. It is in charge of sending the control signal to the centrifugal pump. A data acquisition card is used to read the current state of the plant. The set point signal is the desired liquid level and the process value is the real level measured at the tank. The selected controller is an adaptive PID.

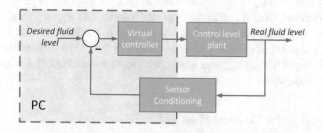

Fig. 2. Control scheme of liquid level plant

A National Instruments data acquisition card (model USB-6008 12-bit 10 KS/s Multifunction I/O) was used to connect the plant and the computer.

2.3   Dataset

During the plant operation, the control signal, the error signal and the three coefficients of the plant transfer function, obtained with Recursive Least Square method, are registered with a sampling frequency of 2 Hz. This five variables will be considered the inputs of the classifier.

To face this study, three different operating points are considered as normal data to train a global the classifier:

- Tank level at 30% and output electric valve closed: 5400 samples.
- Tank level at 50% and output electric valve closed: 5400 samples.
- Tank level at 70% and output electric valve closed: 5400 samples.

Once the classifiers are obtained, they are checked using real generated anomalies, corresponding to the next plant states:

- Tank level at 30% and output electric valve completely open: 5400 samples.
- Tank level at 50% and output electric valve completely open: 5400 samples.
- Tank level at 70% and output electric valve completely open: 5400 samples.

3   Intelligent Techniques Used

The different one-class intelligent techniques used to perform anomaly detection are described in this section.

3.1   Support Vector Machine

The use of Support Vector Machine (SVM) in classification tasks presented successful results in many different application, such as document classification, image retrieval or anomaly detection [8,19].

The aim of this supervised technique is to map the dataset into a high dimensional space using a kernel function. Then, a hyperplane that maximizes the distance between the mapped points and the origin is constructed [28]. The support vectors are defined as the instances placed close to the hyperplane.

After the training process, when a new data arrives to the implemented classifier, it gives the distance from this data to the high dimensional plane. If this distance is positive, the data belongs to the target class and it is considered an anomaly otherwise.

3.2   Approximate Convex Hull

The Approximate Convex Hull (ACH) is a one-class classification that showed very good results in different fields and UCI learning machine repositories [4,5, 10]. This classification technique models the boundaries of an original dataset $S \in \mathbb{R}^n$ using its convex hull. The convex hull $CH(S)$ of a finite set of points is the minimal convex set containing S, and it is defined as the convex combination of points in S according to Eq. 1 [10, 26].

$$CH(S) = \sum_{i=1}^{|S|} \beta x_i \mid \forall i : \beta_i \geq 0 \wedge \sum_{i=1}^{|S|} \beta x_i = 1, x_i \in S \tag{1}$$

The vertexes v of the convex hull can be expanded or contracted from its centroid $c = (1/|S|) \sum_i x_i, \forall x_i \in S$, using a parameter $\lambda \in [0, +\infty)$ following the Eq. 2.

$$v^\lambda : \{\lambda v + (1 - \lambda)c \mid v \in CH(s)\} \tag{2}$$

The main problem of obtaining the convex hull of high dimensional datasets is the high computational cost of this process [4].

To avoid this problem, an alternative method consists of making t random projections original data on $2D$ planes and determine their convex limits on those planes.

According to this approach, when the convex hull approximation is obtained from the training dataset, the criteria to classify a test data is the following: if the data is out of at least one of the t projections, it is considered as an anomaly. Figure 3 shows a $2D$ example where an anomaly point on $\mathbb{R}^3$ space is out of one of the original data projection.

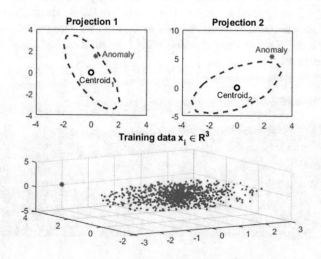

Fig. 3. Anomaly point in $\mathbb{R}^3$

3.3 Artificial Neural Networks. Autoencoder

This widespread technique has been successfully applied in several applications [29], such as denoising data, deep learning or anomaly detection [12, 32].

The one-class classification with Autoencoders uses a nonlinear dimensional reduction with Artificial Neural Network (ANN). The most common supervised learning ANN is the Multilayer Perceptron (MLP), whose performance has proven to be robust [15–17,34].

The main basis of the Autoencoder is to train an ANN that reconstructs the input dataset $x_i \in \mathbb{R}^a$ into the output $\widehat{x}_i$ by an intermediate nonlinear dimensional reduction. Therefore, the number of neurons in the input layer must be equal to the one in the output layer. The basic idea of this method consists of using in the hidden layer at least one less neuron than the input one. This means that the vector $x_i \in \mathbb{R}^a$ is converted to a lower dimensional vector $v_i \in \mathbb{R}^b$.

The input vector dimension reduction leads to remove the abnormal data. After the hidden layer, the information is decompressed and projected at the output, that must replicate the input pattern.

Finally the reconstruction error is calculated as the difference between x_i and $\widehat{x}_i$. If the test data is not consistent with the model, the reconstruction error is high and the data is considered as an anomaly.

4   Experiments and Results

The intelligent techniques described in Sect. 3, were applied to the dataset. According to Subsect. 2.3, three different operating points with the valve closed were considered as normal function. On the other hand, these three operating points with the valve open are considered as anomalies. The techniques were validated using a $k - fold$ cross-validation with $k = 10$. The train and test process are presented in Fig. 4.

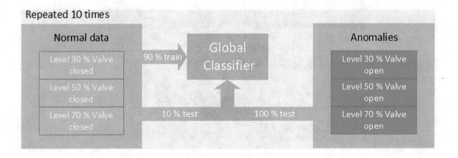

Fig. 4. Train and test process

The Area Under Curve (AUC) parameter (%), is used to assess the performance of each classifier. This indicator relates the false positives and true positives, and it is considered a significant parameter in classification problems [2]. The standard deviation (SD) of the obtained AUC for each $k - fold$ iteration is also calculated.

Due to the fact that the computational must be taken into consideration, the training time (t_{train}) and the latency time (t_{lat}) for each test sample are registered.

The train and test process were performed using a computer with an *Intel Core i7 − 8550U* 1.80 *GHz* processor.

4.1 SVM Classifier

The SVM classifier was implemented using the $'fitcsvm'$ Matlab function [22]. The features were set as follows:

- Kernel function: $'Gaussian'$.
- Outlier fraction: from 0% to 4%.
- Normalization: the input data to the network was normalized in a 0 to 1 interval, or normalized using $Zscore$ method or it was not normalized.
- Once the SVM is trained, the Matlab function $'predict'$ gives the distance to the hyperplane [23]. If this distance is negative, the test data is considered an anomaly.

Table 1 shows the results for each configuration in terms of AUC, SD, t_{train} and t_{lat}.

Table 1. Results obtained with SVM Classifier

Normalization	Outlier frac. (%)	AUC (%)	SD (%)	$t_{train}(s)$	$t_{lat}(\mu s)$
NoNorm	0	72,50	0,04	11,03	75,84
	1	79,98	5,35	10,58	65,25
	2	81,87	1,18	10,43	66,13
	3	82,00	5,84	10,49	66,36
	4	82,42	5,70	10,58	66,33
0 to 1	0	73,42	0,07	10,44	73,26
	1	79,57	6,60	10,43	73,81
	2	81,80	0,92	10,36	73,30
	3	81,63	1,02	10,39	76,99
	4	83,08	5,42	10,43	68,94
Zscore	0	73,19	0,07	10,48	73,96
	1	81,58	6,21	10,49	76,81
	2	81,64	6,20	10,37	76,72
	3	81,88	0,99	11,37	78,03
	4	82,53	4,96	10,38	72,77

4.2   Approximate Convex Hull Classifier

The different parameters were set as follows:

– Parameter λ: it was set to 0,95 (contraction), 1 and 1,05 (expansion).
– Number of $2D$ projections: 50, 100, 500 and 1000.

The results obtained, in terms of AUC, SD and training and testing times are shown in Table 2.

Table 2. Results obtained with ACH Classifier

Parameter λ	Number of proj.	AUC (%)	SD (%)	$t_{train}(s)$	$t_{lat}(ms)$
0,95	50	79,48	0,72	0,07	0,19
	100	77,70	0,53	0,11	0,28
	500	76,86	0,66	0,68	1,86
	1000	76,50	0,65	1,48	4,14
1	50	99,57	0,12	0,05	0,14
	100	99,50	0,14	0,16	0,43
	500	99,38	0,15	0,71	1,96
	1000	99,33	0,14	1,24	3,28
1,05	50	99,98	0,03	0,06	0,16
	100	99,97	0,03	0,12	0,32
	500	99,92	0,06	0,56	1,44
	1000	99,91	0,05	1,63	4,54

4.3   Artificial Neural Network Autoencoder Classifier

The implementation of this technique was done using the Matlab function *'trainAutoencoder'* [21]. The different technique features were configured as follows:

– Number of neurons in the hidden layer: it was set from 1 to $n - 1$, being n the number of inputs. In this case, networks had 1, 2, 3 and 4 neurons in the hidden layer.
– The decision criteria if a new test data does not belong to the target class, follows the reconstruction error criteria. If this value is higher than a specific threshold, the test data is anomalous. To choose a proper threshold, the reconstruction error of the test data is compared with the reconstruction error of the. Then, the percentile of the training data reconstruction error was assessed from 70% to 100%, selecting the percentile threshold that offered better results on the test data.

- The activation function of the hidden layer was set to log-sigmoid, while the output activation function was set to linear.
- Normalization: the dataset to the network was normalized in a 0 to 1 interval, or normalized using $Zscore$ method [31] or it was not normalized.

The values of AUC, SD and training and latency times of this technique are shown in Table 3.

Table 3. Results obtained with Autoencoder Classifier

Normalization	Number of Neurons	AUC (%)	SD (%)	$t_{train}(s)$	$t_{train}(\mu s)$
NoNorm	1	67,77	0,46	275,77	1,09
	2	78,98	0,27	96,32	0,87
	3	95,99	0,51	80,96	0,64
	4	95,97	0,49	127,37	0,65
0 to 1	1	76,80	0,50	1,01	0,76
	2	75,72	0,52	1,35	0,72
	3	74,91	0,40	1,60	0,76
	4	74,25	0,39	1,63	0,79
Zscore	1	66,25	0,53	10,21	0,75
	2	58,36	0,40	10,75	0,82
	3	73,14	2,07	22,48	0,81
	4	93,14	0,53	47,87	0,87

5   Conclusions and Future Works

In this work, a global classifier capable of detect anomalies is proposed. To achieve this goal, three one-class intelligent techniques were applied and evaluated on an a control level plant. Three different operating points with the output valve closed were used as training data to obtain the global classifier, and three with the output valve open were considered as anomalies. Then, approach proposed can detect anomalies over different operating range using only one classifier.

Very good results were obtained, in general terms, specially for the ACH technique. This technique offered successful performance in terms of AUC, training time and latency time. Taking into account these values, the best classifier for this technique were obtained with 50 projections, with an AUC of 99.98%, and extremely low values of t_{train} and t_{lat}.

This work also reflects the importance of λ when defining the convex hull of the ACH classifier. In this case, it can be noticed that a λ value of 0.95 does not offer good results. However, the results obtained for a λ greater or equal than 1 offer better AUC values.

The results of Autoencoder classifiers present significant variations in terms of AUC depending on the number of neurons in the hidden layer. In the case under study, an increase in the number of neurons leads to better results. It can be remarked that the normalization type has strong influence in the training time and AUC obtained.

Finally, the SVM classifiers performance are not as good as the performances of the other techniques, obtaining an AUC value of 83.08% in the best case.

The proposed approach could be implemented over many industrial plants using data from their right operation. Then, the appearance of anomalies can be detected, improving the system performance and the maintenance planning, among others. Also, the use of a global classifier can guarantee the anomaly detection in different operating ranges.

As future works, the use of local classifiers for each operating point can be considered as future works. Furthermore, using Dimensional Reduction Techniques (DRT) [27,30] can be applied to reduce the computational cost of the techniques proposed.

Due to the fact that the system evolves with its use, the initial dataset may not be representative after a certain time. Then, the possibility of retrain the classifiers with new data can be taken into consideration.

References

1. Baruque, B., Porras, S., Jove, E., Calvo-Rolle, J.L.: Geothermal heat exchanger energy prediction based on time series and monitoring sensors optimization. Energy **171**, 49–60 (2019)
2. Bradley, A.P.: The use of the area under the ROC curve in the evaluation of machine learning algorithms. Pattern Recognit. **30**(7), 1145–1159 (1997)
3. Calvo-Rolle, J.L., Quintian-Pardo, H., Corchado, E., del Carmen Meizoso-López, M., García, R.F.: Simplified method based on an intelligent model to obtain the extinction angle of the current for a single-phase half wave controlled rectifier with resistive and inductive load. J. Appl. Logic **13**(1), 37–47 (2015)
4. Casale, P., Pujol, O., Radeva, P.: Approximate convex hulls family for one-class classification. In: International Workshop on Multiple Classifier Systems, pp. 106–115. Springer (2011)
5. Casale, P., Pujol, O., Radeva, P.: Approximate convex hulls family for one-class classification. In: Sansone, C., Kittler, J., Roli, F. (eds.) Multiple Classifier Systems, pp. 106–115. Springer, Heidelberg (2011)
6. Casteleiro-Roca, J.L., Jove, E., Gonzalez-Cava, J.M., Méndez Pérez, J.A., Calvo-Rolle, J.L., Blanco Alvarez, F.: Hybrid model for the ani index prediction using remifentanil drug and emg signal. Neural Comput. Appl. (2018). https://doi.org/10.1007/s00521-018-3605-z
7. Chandola, V., Banerjee, A., Kumar, V.: Anomaly detection: a survey. ACM Comput. Surv. (CSUR) **41**(3), 15 (2009)
8. Chen, Y., Zhou, X.S., Huang, T.S.: One-class SVM for learning in image retrieval. In: Proceedings of the 2001 International Conference on Image Processing, vol. 1, pp. 34–37. IEEE (2001)
9. Chiang, L.H., Russell, E.L., Braatz, R.D.: Fault Detection and Diagnosis in Industrial Systems. Springer, London (2000)

10. Fernández-Francos, D., Fontenla-Romero, Ó., Alonso-Betanzos, A.: One-class convex hull-based algorithm for classification in distributed environments. IEEE Trans. Syst. Man Cybernet. Syst. 1–11 (2018)
11. González, G., Angelo, C.D., Forchetti, D., Aligia, D.: Diagnóstico de fallas en el convertidor del rotor en generadores de inducción con rotor bobinado. Revista Iberoamericana de Automática e Informática industrial **15**(3), 297–308 (2018). https://polipapers.upv.es/index.php/RIAI/article/view/9042
12. Goodfellow, I., Bengio, Y., Courville, A., Bengio, Y.: Deep Learning, vol. 1. MIT Press, Cambridge (2016)
13. Hobday, M.: Product complexity, innovation and industrial organisation. Res. Policy **26**(6), 689–710 (1998)
14. Hodge, V., Austin, J.: A survey of outlier detection methodologies. Artif. Intell. Rev. **22**(2), 85–126 (2004)
15. Jove, E., Aláiz-Moretón, H., Casteleiro-Roca, J.L., Corchado, E., Calvo-Rolle, J.L.: Modeling of bicomponent mixing system used in the manufacture of wind generator blades. In: Corchado, E., Lozano, J.A., Quintián, H., Yin, H. (eds.) Intelligent Data Engineering and Automated Learning - IDEAL 2014, pp. 275–285. Springer International Publishing, Cham (2014)
16. Jove, E., Antonio Lopez-Vazquez, J., Isabel Fernandez-Ibanez, M., Casteleiro-Roca, J.L., Luis Calvo-Rolle, J.: Hybrid intelligent system to predict the individual academic performance of engineering students. Int. J. Eng. Educ. **34**(3), 895–904 (2018)
17. Jove, E., Gonzalez-Cava, J.M., Casteleiro-Roca, J.L., Méndez-Pérez, J.A., Antonio Reboso-Morales, J., Javier Pérez-Castelo, F., Javier de Cos Juez, F., Luis Calvo-Rolle, J.: Modelling the hypnotic patient response in general anaesthesia using intelligent models. Logic J. IGPL (2018)
18. Moreno-Fernandez-de Leceta, A., Lopez-Guede, J.M., Ezquerro Insagurbe, L., Ruiz de Arbulo, N., Granã, M.: A novel methodology for clinical semantic annotations assessment. Logic J. IGPL **26**(6), 569–580 (2018). http://dx.doi.org/10.1093/jigpal/jzy021
19. Li, K.L., Huang, H.K., Tian, S.F., Xu, W.: Improving one-class SVM for anomaly detection. In: 2003 International Conference on Machine Learning and Cybernetics, vol. 5, pp. 3077–3081. IEEE (2003)
20. Manuel Vilar-Martinez, X., Aurelio Montero-Sousa, J., Luis Calvo-Rolle, J., Luis Casteleiro-Roca, J.: Expert system development to assist on the verification of "tacan" system performance. Dyna **89**(1), 112–121 (2014)
21. MathWorks: Autoencoder, 29 January 2019. https://es.mathworks.com/help/deeplearning/ref/trainautoencoder.html
22. MathWorks: fitcsvm, 29 January 2019. https://es.mathworks.com/help/stats/fitcsvm.html
23. MathWorks: predict, 29 January 2019. https://es.mathworks.com/help/stats/classreg.learning.classif.compactclassificationsvm.predict.html
24. Miljković, D.: Fault detection methods: a literature survey. In: MIPRO, 2011 Proceedings of the 34th International Convention, pp. 750–755. IEEE (2011)
25. de la Portilla, M.P., Eiro, A.L.P., Sánchez, J.A.S., Herrera, R.M.: Modelado dinámico y control de un dispositivo sumergido provisto de actuadores hidrostáticos. Revista Iberoamericana de Automática e Informática industrial **15**(1), 12–23 (2017). https://polipapers.upv.es/index.php/RIAI/article/view/8824
26. Preparata, F.P., Shamos, M.I.: Computational Geometry: An Introduction. Springer, New York (2012)

27. Quintián, H., Corchado, E.: Beta scale invariant map. Eng. Appl. Artif. Intell. **59**, 218–235 (2017)
28. Rebentrost, P., Mohseni, M., Lloyd, S.: Quantum support vector machine for big data classification. Phys. Rev. Lett. **113**, 130503 (2014). https://link.aps.org/doi/10.1103/PhysRevLett.113.130503
29. Sakurada, M., Yairi, T.: Anomaly detection using autoencoders with nonlinear dimensionality reduction. In: Proceedings of the MLSDA 2014 2nd Workshop on Machine Learning for Sensory Data Analysis, p. 4. ACM (2014)
30. Segovia, F., Górriz, J.M., Ramírez, J., Martinez-Murcia, F.J., García-Pérez, M.: Using deep neural networks along with dimensionality reduction techniques to assist the diagnosis of neurodegenerative disorders. Logic J. IGPL **26**(6), 618–628 (2018). http://dx.doi.org/10.1093/jigpal/jzy026
31. Shalabi, L.A., Shaaban, Z.: Normalization as a preprocessing engine for data mining and the approach of preference matrix. In: 2006 International Conference on Dependability of Computer Systems, pp. 207–214, May 2006
32. Vincent, P., Larochelle, H., Lajoie, I., Bengio, Y., Manzagol, P.A.: Stacked denoising autoencoders: learning useful representations in a deep network with a local denoising criterion. J. Mach. Learn. Res. **11**(Dec), 3371–3408 (2010)
33. Wojciechowski, S.: A comparison of classification strategies in rule-based classifiers. Logic J. IGPL **26**(1), 29–46 (2018). http://dx.doi.org/10.1093/jigpal/jzx053
34. Zeng, Z., Wang, J.: Advances in Neural Network Research and Applications, 1st edn. Springer, Heidelberg (2010)

Doctoral Consortium

Organization of Doctoral Consortium Sessions

The aim of the Doctoral Consortium is to provide a frame where students can present their ongoing research work and meet other students and researchers, and obtain feedback on future research directions. The Doctoral Consortium is intended for students who have a specific research proposal and some preliminary results, but who are still far from completing their dissertation.

All proposals submitted to the Doctoral Consortium were undergo a thorough reviewing process with the aim to provide detailed and constructive feedback.

The submissions should identify:

- Problem statement
- Related work
- Hypothesis
- Proposal
- Preliminary Results and/or Evaluation Plan
- Reflections

Organization

Doctoral Consortium Organizer

Sara Rodríguez   University of Salamanca, Spain

Technical Evaluation of Plug-in Electric Vehicles Charging Load on a Real Distribution Grid

Behzad Hashemi[1] and Payam Teimourzadeh Baboli[2(✉)]

[1] Babol Noshirvani University of Technology, Babol, Iran
b.hashemi@stu.nit.ac.ir
[2] University of Mazandaran, Babolsar, Iran
p.teimourzadeh@umz.ac.ir

Abstract. The popularity of Plug-in Electric Vehicles (PEVs) in the last few years however is a turning point toward alleviating the global warming, but the inevitable effects of charging load of these vehicles on electric grids has become a concern for grid operators. While uncoordinated charging of a large number of PEVs may jeopardize the operation of the grids, intelligent methods can be used to coordinate the charging processes for the benefit of the grids. This paper presents a comprehensive model of future charging load of PEVs in a real distribution grid by considering PEVs' characteristics and different driving patterns. Domestic and public charging are both considered. Moreover, an intelligent approach based on Non-dominated Sorting Genetic Algorithm (NSGA-II) will be introduced to coordinate PEVs' charging with the aim of minimizing the power losses cost of the grid and maximizing the PEV owners' satisfaction and considering technical constraints in our next work. This study is carried out on a real medium voltage distribution grid of Tehran Province Distribution Company in Lavasan city in Iran. The results show the detrimental effects of uncoordinated charging of PEVs on the operation of the grid which can be reduced by implementing the mentioned intelligent coordination approach.

Keywords: Plug-in Electric Vehicle · NSGA-II · Distribution grid

1 Introduction

The market share of Plug-in Electric Vehicles (PEVs) is rising all over the world. This is of significance importance for grid operators due to direct connection of PEVs into distribution grids. This necessitates evaluating the charging load of these vehicles on the grid. Many papers has studied different financial and technical aspects of grid operation with the high presence of PEVs. Increasing the peak demand load of feeders, power losses and operation costs of grids, and voltage deviation of nodes are among the most important ones [1]. Therefore, various techniques with different objectives have been proposed in the literature to benefit from PEVs. Utilization of Classical optimization methods in [2–16] together with using heuristic methods in [17] are examples in different field science such as efficient scheduling of these vehicles [18–68].

© Springer Nature Switzerland AG 2020
E. Herrera-Viedma et al. (Eds.): DCAI 2019, AISC 1004, pp. 163–170, 2020.
https://doi.org/10.1007/978-3-030-23946-6_18

This paper investigates technical impact of widespread presence of PEVs in a real medium voltage distribution grid in Iran. The stochastic driving patterns are also modeled considering domestic and public charging to obtain reliable results. Moreover, a new multiobjective approach based on NSGA-II will be proposed to minimize power losses cost of the grid and maximize the PEV owners' satisfaction as our next work.

2 Problem Formulation

2.1 Modeling of Driving Patterns and PEVs

Driving patterns are usually introduced by three attributes including arrival time, departure time, and daily mileage. According to real-world data, two first attributes and the last one can be generated using normal and lognormal probability density functions (pdfs), respectively. In addition, two types of charging are considered: domestic and public (office and shopping). The parameters of the related pdfs are presented in Table 1. The driving attributes are randomly generated for every vehicle.

Table 1. Parameters of driving patterns model.

	Home parking		Office parking		Shopping parking		Daily Mileage
	Arr. Time	Dep. Time	Arr. Time	Dep. Time	Arr. Time	Dep. Time	
Mean	20	7.5	8.5	14	16	18	3.715
Standard deviation	1.5	0.75	0.5	0.5	1	1	0.6

PEVs are modeled based on their batteries. State of Charge (SoC) of a battery is the ratio of the energy stored in the battery to its capacity. Moreover, SoC is directly affected by the energy absorbed from the grid and the energy consumed on the road.

2.2 Mathematical Formulation of the Coordinated Charging Mode

As mentioned earlier, an NSGA-II based algorithm will be implemented to schedule charging processes of PEVs. The objective function of this optimization problem will be to minimize daily power losses cost of the grid and maximize the total energy absorbed by PEVs in a day as an index of PEV owners' satisfaction. The formulation is as follows.

$$Objective : \text{Minimize} \left(\text{Cost}^{power\,losses} \right) \, \& \, \text{Maximize} \left(\text{Power}^{customers} \right) \quad (1)$$

In addition, the constraints of grid operation including current flow though lines and voltage deviation of nodes and the technical constraints of SoC limitations will be considered in the problem. Finally, an NSGA-II algorithm will be used to solve the proposed problem. NSGA-II is well-known for its accuracy and computational capabilities which can be used to solve multi-objective optimization problems.

3   Case Study and First Results

A real 20 kV 156 node distribution grid of Tehran Province Distribution Company named Saheli in the city of Lavasan (Iran) is modeled in MATLAB. The active and reactive power of the main feeder on the 24th of July 2017 has been chosen as base load. Moreover, 8 different types of PEVs with different characteristics has been chosen. The study is carried out on 3 penetration levels of PEVs including 10%, 30%, and 50%. The main results of uncoordinated charging mode is shown in Fig. 1.

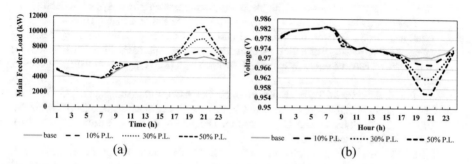

Fig. 1.   (a) Daily main feeder load curve and (b) daily voltage curve of the last node, for different penetration levels.

As can be seen, uncoordinated charging of PEVs increases the peak demand load of the feeder and voltage deviation of the nodes which both get worse by increasing the penetration level. Daily power loss is also increased from 2299 kW in base load to 3138 kW in 50% penetration level. The proposed coordinated mode is under study to improve operation indices of the grid while PEV owners' satisfaction is not violated.

References

1. Shafiee, S., Fotuhi-Firuzabad, M., Rastegar, M.: Investigating the impacts of plug-in hybrid electric vehicles on power distribution systems. IEEE Trans. Smart Grid **4**, 1351–1360 (2013)
2. Hafez, O., Bhattacharya, K.: Queuing analysis based PEV load modeling considering battery charging behavior and their impact on distribution system operation. IEEE Trans. Smart Grid **9**(1), 261–273 (2018)
3. Hashemi, B., Shahabi, M., Teimourzadeh Baboli, P.: Stochastic based optimal charging strategy for plug-in electric vehicles aggregator under incentive and regulatory policies of DSO. IEEE Trans. Veh. Technol. pp. 1–11 (2019)
4. Gazafroudi, A.S., Corchado, J.M., Kean, A., Soroudi, A.: Decentralized flexibility management for electric vehicles. IET Renew. Power Gener. (2019). http://ietdl.org/t/IBgIPb
5. Gazafroudi, A.S., Soares, J., Ghazvini, M.A.F., Pinto, T., Vale, Z., Corchado, J.M.: Stochastic interval-based optimal offering model for residential energy management systems by household owners. Int. J. Electr. Power Energy Syst. **105**, 201–219 (2019)

6. Prieto-Castrillo, F., Shokri Gazafroudi, A., Prieto, J., Corchado, J.M.: An Ising spin-based model to explore efficient flexibility in distributed power systems. Complexity (2018)

7. Gazafroudi, A.S., Prieto-Castrillo, F., Pinto, T., Prieto, J., Corchado, J.M., Bajo, J.: Energy flexibility management based on predictive dispatch model of domestic energy management system. Energies 10(9), 1397 (2017)

8. Gazafroudi, A.S., Shafie-Khah, M., Abedi, M., Hosseinian, S.H., Dehkordi, G.H., Goel, L., Karimyan, P., Prieto-Castrillo, F., Corchado, J.M., Catalão, J.P.: A novel stochastic reserve cost allocation approach of electricity market agents in the restructured power systems. Electr. Power Syst. Res. 152, 223–236 (2017)

9. Gazafroudi, A.S., Shafie-khah, M., Fitiwi, D.Z., Santos, S.F., Corchado, J.M., Catalão, J.P.: Impact of strategic behaviors of the electricity consumers on power system reliability. In: Sustainable Interdependent Networks II, pp. 193–215. Springer, Cham (2019)

10. Gazafroudi, A.S., Prieto-Castrillo, F., Pinto, T., Corchado, J.M.: Energy flexibility management in power distribution systems: decentralized approach. In: 2018 International Conference on Smart Energy Systems and Technologies (SEST), pp. 1–6. IEEE, September 2018

11. Gazafroudi, A.S., Pinto, T., Prieto-Castrillo, F., Corchado, J.M., Abrishambaf, O., Jozi, A., Vale, Z.: Energy flexibility assessment of a multi agent-based smart home energy system. In: 2017 IEEE 17th International Conference on Ubiquitous Wireless Broadband (ICUWB), pp. 1–7. IEEE, September 2017

12. Bajool, R., Shafie-khah, M., Gazafroudi, A.S., Catalão, J.P.: Mitigation of active and reactive demand response mismatches through reactive power control considering static load modeling in distribution grids. In: 2017 IEEE Conference on Control Technology and Applications (CCTA), pp. 1637–1642. IEEE, August 2017

13. Gazafroudi, A.S., Prieto-Castrillo, F., Pinto, T., Jozi, A., Vale, Z.: Economic evaluation of predictive dispatch model in MAS-based smart home. In: International Conference on Practical Applications of Agents and Multi-Agent Systems, pp. 81–91. Springer, Cham, June 2017

14. Gazafroudi, A.S., De Paz, J.F., Prieto-Castrillo, F., Villarrubia, G., Talari, S., Shafie-khah, M., Catalão, J.P.: A review of multi-agent based energy management systems. In: International Symposium on Ambient Intelligence, pp. 203–209. Springer, Cham, June 2017

15. Gazafroudi, A.S., Pinto, T., Prieto-Castrillo, F., Prieto, J., Corchado, J.M., Jozi, A., Vale, Z., Venayagamoorthy, G.K.: Organization-based multi-agent structure of the smart home electricity system. In: 2017 IEEE Congress on Evolutionary Computation (CEC), pp. 1327–1334. IEEE, June 2017

16. Gazafroudi, A.S., Prieto-Castrillo, F., Corchado, J.M.: Residential energy management using a novel interval optimization method. In: 2017 4th International Conference on Control, Decision and Information Technologies (CoDIT), pp. 0196–0201. IEEE April 2017

17. Gong, L., Cao, W., Liu, K., Zhao, J., Li, X.: Spatial and temporal optimization strategy for plug-in electric vehicle charging to mitigate impacts on distribution network. Energies 11(6), 1373 (2018)

18. Kang, Q., Feng, S., Zhou, M., Ammari, A.C., Sedraoui, K.: Optimal load scheduling of plug-in hybrid electric vehicles via weight aggregation multi-objective evolutionary algorithms. IEEE Trans. Intell. Transp. Syst. 18(9), 2557–2568 (2017)

19. Najafi, S., Talari, S., Gazafroudi, A.S., Shafie-khah, M., Corchado, J.M., Catalão, J.P.: Decentralized control of DR using a multi-agent method. In: Sustainable Interdependent Networks, pp. 233–249. Springer, Cham (2018)

20. Ebrahimi, M., Gazafroudi, A. S., Corchado, J. M., Ebrahimi, M.: Energy management of smart home considering residences' satisfaction and PHEV. In: 2018 International Conference on Smart Energy Systems and Technologies (SEST), pp. 1–6. IEEE September 2018

21. Pinto, T., Gazafroudi, A.S., Prieto-Castrillo, F., Santos, G., Silva, F., Corchado, J.M., Vale, Z.: Reserve costs allocation model for energy and reserve market simulation. In: 2017 19th International Conference on Intelligent System Application to Power Systems (ISAP), pp. 1–6. IEEE, September 2017

22. Navarro-Cáceres, M., Gazafroudi, A.S., Prieto-Castillo, F., Venyagamoorthy, K.G., Corchado, J.M.: Application of artificial immune system to domestic energy management problem. In: 2017 IEEE 17th International Conference on Ubiquitous Wireless Broadband (ICUWB), pp. 1–7. IEEE, September 2017

23. Hernández, E., González, A., Pérez, B., de Luis Reboredo, A., Rodríguez, S.: Virtual organization for fintech management. In: International Symposium on Distributed Computing and Artificial Intelligence, pp. 201–210. Springer, Cham, June 2018

24. Hernández, E., Sittón, I., Rodríguez, S., Gil, A.B., García, R.J.: An investment recommender multi-agent system in financial technology. In: The 13th International Conference on Soft Computing Models in Industrial and Environmental Applications, pp. 3–10. Springer, Cham, June 2018

25. Candanedo, I.S., Nieves, E. H., González, S.R., Martín, M.T.S., Briones, A.G.: Machine learning predictive model for industry 4.0. In: International Conference on Knowledge Management in Organizations, pp. 501–510. Springer, Cham, August 2018

26. Nassaj, A., Shahrtash, S.M.: An accelerated preventive agent based scheme for post-disturbance voltage control and loss reduction. IEEE Trans. Power Syst. 33(44), 4508–4518 (2018)

27. Nassaj, A., Shahrtash, S.M.: A predictive agent-based scheme for post-disturbance voltage control. Int. J. Electr. Power Energy Syst. 98, 189–198 (2018)

28. Casado-Vara, R., Chamoso, P., De la Prieta, F., Prieto J., Corchado J.M.: Non-linear adaptive closed-loop control system for improved efficiency in IoT-blockchain management. Inf. Fusion (2019)

29. González-Briones, A., Chamoso, P., Yoe, H., Corchado, J.M.: GreenVMAS: virtual organization based platform for heating greenhouses using waste energy from power plants. Sensors 18(3), 861 (2018)

30. Casado-Vara, R., Novais, P., Gil, A.B., Prieto, J., Corchado, J.M.: Distributed continuous-time fault estimation control for multiple devices in IoT networks. IEEE Access (2019)

31. Chamoso, P., González-Briones, A., Rivas, A., De La Prieta, F., Corchado J.M.: Social computing in currency exchange. Knowl. Inf. Syst. (2019)

32. Casado-Vara, R., Prieto-Castrillo, F., Corchado, J.M.: A game theory approach for cooperative control to improve data quality and false data detection in WSN. Int. J. Robust Nonlinear Control 28(16), 5087–5102 (2018)

33. Morente-Molinera, J.A., Kou, G., González-Crespo, R., Corchado, J.M., Herrera-Viedma, E.: Solving multi-criteria group decision making problems under environments with a high number of alternatives using fuzzy ontologies and multi-granular linguistic modelling methods. Knowl.-Based Syst. 137, 54–64 (2017)

34. Li, T., Sun, S., Bolić, M., Corchado, J.M.: Algorithm design for parallel implementation of the SMC-PHD filter. Sig. Process. 119, 115–127 (2016). https://doi.org/10.1016/j.sigpro.2015.07.013

35. Chamoso, P., Rodríguez, S., de la Prieta, F., Bajo, J.: Classification of retinal vessels using a collaborative agent-based architecture. AI Commun. (Preprint), 1–18 (2018)

36. Chamoso, P., González-Briones, A., Rodríguez, S., Corchado, J.M.: Tendencies of technologies and platforms in smart cities: a state-of-the-art review. Wirel. Commun. Mob. Comput. (2018)
37. Gonzalez-Briones, A., Prieto, J., De La Prieta, F., Herrera-Viedma, E., Corchado, J.M.: Energy optimization using a case-based reasoning strategy. Sensors (Basel) **18**(3), 865–865 (2018). https://doi.org/10.3390/s18030865
38. Gonzalez-Briones, A., Chamoso, P., De La Prieta, F., Demazeau, Y., Corchado, J.M.: Agreement technologies for energy optimization at home. Sensors (Basel) **18**(5), 1633 (2018). https://doi.org/10.3390/s18051633
39. Di Mascio, T., Vittorini, P., Gennari, R., Melonio, A., De La Prieta, F., Alrifai, M.: The learners' user classes in the TERENCE adaptive learning system. In: 2012 IEEE 12th International Conference on Advanced Learning Technologies, pp. 572–576. IEEE
40. Tapia, D.I., Alonso, R.S., De Paz, J.F., Zato, C., Prieta, F.D.L.: A telemonitoring system for healthcare using heterogeneous wireless sensor networks. Int. J. Artif. Intell. **6**(S11), 112–128 (2011)
41. de la Prieta, F., Navarro, M., García, J.A., González, R., Rodríguez, S.: Multi-agent system for controlling a cloud computing environment. In: Portuguese Conference on Artificial Intelligence, pp. 13–20. Springer, Heidelberg, September 2013
42. Chamoso, P., Rivas, A., Martín-Limorti, J.J., Rodríguez, S.: A hash based image matching algorithm for social networks. In: Advances in Intelligent Systems and Computing, vol. 619, pp. 183–190 (2018). https://doi.org/10.1007/978-3-319-61578-3_18
43. Sittón, I., Rodríguez, S.: Pattern extraction for the design of predictive models in industry 4.0. In: International Conference on Practical Applications of Agents and Multi-Agent Systems, pp. 258–261 (2017)
44. García, O., Chamoso, P., Prieto, J., Rodríguez, S., De La Prieta, F.: A serious game to reduce consumption in smart buildings. In: Communications in Computer and Information Science vol. 722, pp. 481–493 (2017). https://doi.org/10.1007/978-3-319-60285-1_41
45. Palomino, C.G., Nunes, C.S., Silveira, R.A., González, S.R., Nakayama, M.K.: Adaptive agent-based environment model to enable the teacher to create an adaptive class. In: Advances in Intelligent Systems and Computing, vol. 617 (2017). https://doi.org/10.1007/978-3-319-60819-8_3
46. Canizes, B., Pinto, T., Soares, J., Vale, Z., Chamoso, P., Santos, D.: Smart City: A GECAD-BISITE energy management case study. In: 15th International Conference on Practical Applications of Agents and Multi-Agent Systems PAAMS 2017, Trends in Cyber-Physical Multi-Agent Systems, vol. 2, pp. 92–100 (2017). https://doi.org/10.1007/978-3-319-61578-3_9
47. Chamoso, P., de La Prieta, F., Eibenstein, A., Santos-Santos, D., Tizio, A., Vittorini, P.: A device supporting the self management of tinnitus. In: Lecture Notes in Computer Science (including subseries Lecture Notes in Artificial Intelligence and Lecture Notes in Bioinformatics). LNCS, vol. 10209, pp. 399–410 (2017). https://doi.org/10.1007/978-3-319-56154-7_36
48. Román, J.A., Rodríguez, S., de da Prieta, F.: Improving the distribution of services in MAS. In: Communications in Computer and Information Science, vol. 616 (2016). https://doi.org/10.1007/978-3-319-39387-2_4
49. Buciarelli, E., Silvestri, M., González, S.R.: Decision economics. In: 13th International Conference on Commemoration of the Birth Centennial of Herbert A. Simon 1916–2016 (Nobel Prize in Economics 1978): Distributed Computing and Artificial Intelligence. Advances in Intelligent Systems and Computing, vol. 475. Springer (2016)

50. Lima, A.C.E.S., De Castro, L.N., Corchado, J.M.: A polarity analysis framework for Twitter messages. Appl. Math. Comput. **270**, 756–767 (2015). https://doi.org/10.1016/j.amc.2015.08.059

51. Redondo-Gonzalez, E., De Castro, L.N., Moreno-Sierra, J., Maestro De Las Casas, M.L., Vera-Gonzalez, V., Ferrari, D.G., Corchado, J.M.: Bladder carcinoma data with clinical risk factors and molecular markers: a cluster analysis. BioMed. Res. Int. (2015). https://doi.org/10.1155/2015/168682

52. Li, T., Sun, S., Corchado, J. M., Siyau, M.F.: Random finite set-based Bayesian filters using magnitude-adaptive target birth intensity. In: FUSION 2014 - 17th International Conference on Information Fusion (2014). https://www.scopus.com/inward/record.uri?eid=2-s2.0-84910637788&partnerID=40&md5=bd8602d6146b014266cf07dc35a681e0

53. Prieto, J., Alonso, A.A., de la Rosa, R., Carrera, A.: Adaptive framework for uncertainty analysis in electromagnetic field measurements. Radiat. Prot. Dosimetry, ncu260 (2014)

54. Chamoso, P., Raveane, W., Parra, V., González, A.: Uavs applied to the counting and monitoring of animals. In: Advances in Intelligent Systems and Computing, vol. 291, pp. 71–80 (2014). https://doi.org/10.1007/978-3-319-07596-9_8

55. Pérez, A., Chamoso, P., Parra, V., Sánchez, A.J.: Ground vehicle detection through aerial images taken by a UAV. In: 2014 17th International Conference on Information Fusion (FUSION) (2014)

56. Choon, Y.W., Mohamad, M.S., Deris, S., Illias, R.M., Chong, C.K., Chai, L.E., Omatu, S., Corchado, J.M.: Differential bees flux balance analysis with OptKnock for in silico microbial strains optimization. PLoS ONE 9(7) (2014). https://doi.org/10.1371/journal.pone.0102744

57. Li, T., Sun, S., Corchado, J. M., Siyau, M.F.: A particle dyeing approach for track continuity for the SMC-PHD filter. In: FUSION 2014 - 17th International Conference on Information Fusion (2014). https://www.scopus.com/inward/record.uri?eid=2-s2.0-84910637583&partnerID=40&md5=709eb4815eaf544ce01a2c21aa749d8f

58. García Coria, J.A., Castellanos-Garzón, J.A., Corchado, J.M.: Intelligent business processes composition based on multi-agent systems. Expert Syst. Appl. **41**(4 PART 1), 1189–1205 (2014). https://doi.org/10.1016/j.eswa.2013.08.003

59. Heras, S., De la Prieta, F., Julian, V., Rodríguez, S., Botti, V., Bajo, J., Corchado, J.M.: Agreement technologies and their use in cloud computing environments. Prog. Artif. Intell. **1**(4), 277–290 (2012)

60. Prieto, J., Mazuelas, S., Bahillo, A., Fernández, P., Lorenzo, R. M., Abril, E.J.: Accurate and robust localization in harsh environments based on V2I communication. In: Vehicular Technologies - Deployment and Applications. INTECH Open Access Publisher (2013)

61. Tapia, D.I., Fraile, J.A., Rodríguez, S., Alonso, R.S., Corchado, J.M.: Integrating hardware agents into an enhanced multi-agent architecture for ambient intelligence systems. Inf. Sci. **222**, 47–65 (2013). https://doi.org/10.1016/j.ins.2011.05.002

62. Prieto, J., Mazuelas, S., Bahillo, A., Fernandez, P., Lorenzo, R.M., Abril, E.J.: Adaptive data fusion for wireless localization in harsh environments. IEEE Trans. Signal Process. **60**(4), 1585–1596 (2012)

63. Muñoz, M., Rodríguez, M., Rodríguez, M.E., Rodríguez, S.: Genetic evaluation of the class III dentofacial in rural and urban Spanish population by AI techniques. In: Advances in Intelligent and Soft Computing, vol. 151. AISC (2012). https://doi.org/10.1007/978-3-642-28765-7_49

64. Costa, Â., Novais, P., Corchado, J.M., Neves, J.: Increased performance and better patient attendance in an hospital with the use of smart agendas. Logic J. IGPL **20**(4), 689–698 (2012). https://doi.org/10.1093/jigpal/jzr021

65. García, E., Rodríguez, S., Martín, B., Zato, C., Pérez, B.: MISIA: middleware infrastructure to simulate intelligent agents. In: Advances in Intelligent and Soft Computing, vol. 91 (2011). https://doi.org/10.1007/978-3-642-19934-9_14

66. Rodríguez, S., De La Prieta, F., Tapia, D.I., Corchado, J.M.: Agents and computer vision for processing stereoscopic images. In: Lecture Notes in Computer Science (including subseries Lecture Notes in Artificial Intelligence and Lecture Notes in Bioinformatics), LNAI. vol. 6077 (2010). https://doi.org/10.1007/978-3-642-13803-4_12

67. Rodríguez, S., Gil, O., De La Prieta, F., Zato, C., Corchado, J.M., Vega, P., Francisco, M.: People detection and stereoscopic analysis using MAS. In: INES 2010 - 14th International Conference on Intelligent Engineering Systems, Proceedings (2010) https://doi.org/10.1109/INES.2010.5483855

68. Prieto, J., Mazuelas, S., Bahillo, A., Fernández, P., Lorenzo, R.M., Abril, E.J.:. On the minimization of different sources of error for an RTT-based indoor localization system without any calibration stage. In: 2010 International Conference on Indoor Positioning and Indoor Navigation (IPIN), pp. 1–6 (2010)

An Agent-Based Approach for Market-Based Customer Reliability Enhancement in Distribution Systems

Mahan Ebrahimi[1]($\boxtimes$), Mahoor Ebrahimi[2], and Behzad Abdi[1]

[1] Sharif University of Technology, Tehran, Iran
Ebrahimi.mahan@yahoo.com
[2] Amirkabir University of Technology, Tehran, Iran

Abstract. These days, considering the importance of reliability in industries, utilities' role for preparing high levels of reliability becomes more important. This can be done by adding numbers of switches and Tie switches which impose high expenses to system. These expenditures should be provided by customers. Level of reliability in different points of network is different. So customers which pay more, receive higher level of reliability. If a customer raises its payment, utility change the location of switches in order to raise level of reliability of that customer. all customers change their payment and then utility fixes the switches at a point that all customers reach their favorable reliability. Moreover, there are some customers called "free riders". They receive high levels of reliability because of their closeness to the customers who pay much and utility provide them high level of reliability. So free riders get high level of reliability without paying much. Therefore, a solution is needed for the mentioned problems.

Keywords: Distribution system · Reliability market · Multi-agent systems

1 Proposed Scheme

Reliability enhancement and reducing Costs are two goals of power distribution utilities' costumers [1–7]. There are so many costumers which need specific level of reliability and utilities prepare these levels of reliability by adding some switches or changing their location based on costumers' payments. Usually distribution system configuration is a loop but it can be used in radial form by using some switches. So all customers can receive power from at least two Transformers. There are different kind of switches such as remote control switch, manual switches, recloser and tie switches. The difference between these kind of switches is their activation time. Thus, the costumers which are near to the kinds of switches and have low activation time, they have short restoration time. Moreover, in the sections of network that the distance between switches is low, the probability of fault occurrence is smaller than other parts.

In this way, an independent distribution system operator (DSO) is needed. This system is in charge of minimizing costs and reaching maximum reliability based on customers' proposes. Another system in proposed model is Customers. Customers need

© Springer Nature Switzerland AG 2020
E. Herrera-Viedma et al. (Eds.): DCAI 2019, AISC 1004, pp. 171–176, 2020.
https://doi.org/10.1007/978-3-030-23946-6_19

different levels of reliability. Reliability is evaluated by different indices such as SAIDI (system average interruption duration index) and SAIFI (system average interruption frequency index). Each outage imposes huge costs to customers. These costs are different for different customers. Therefore, each of them define a damage cost function [8].

Customers are in a competition for reaching to their desirable SAIDI. They propose their bid for payment to DSO and DSO finds optimum number and location for switches. some parts of payments assign for buying switches and other parts allocate for changing the location of switches based on their cost function. In this way, customers act as agents in the system. Multi-agent approach is practical in different fields of science [9–42] such as power systems [43–59]. The operating system of proposed model is switches. They receive command from DSO; make favorable changes and SAIDI of all customers changes. So we can consider new SAIDIs as an output from switches to Costumers.

Also, a communication system is needed for sending data such as bids and damage cost functions from customers to DSO. Other DSO's inputs are feeder's active and reactive powers. Figure 1 shows relation between systems; DSO, costumers and switches.

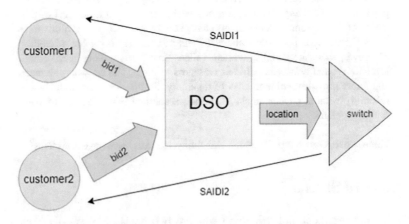

Fig. 1. General structure for market-based distribution system

References

1. Brown, R.E.: Impact of smart grid on distribution system design. In: IEEE Power Energy Society General Meeting-Conversion and Delivery of Electrical Energy in the 21st Century, PES, pp. 2008–2011 (2008)
2. Ramandi, M.Y., Afshar, K., Gazafroudi, A.S., Bigdeli, N.: Reliability and economic evaluation of demand side management programming in wind integrated power systems. Int. J. Electr. Power Energy Syst. **78**, 258–268 (2016)
3. Gazafroudi, A.S., Afshar, K., Bigdeli, N.: Assessing the operating re-serves and costs with considering customer choice and wind power uncertainty in pool-based power market. Int. J. Electr. Power Energy Syst. **67**, 202–215 (2015)

4. Afshar, K., Shokri Gazafroudi, A.: Application of stochastic programming to determine operating reserves with considering wind and load uncertainties. J. Oper. Autom. Power Eng. **1**(2), 96–109 (2007)
5. Gazafroudi, A.S., Shafie-Khah, M., Abedi, M., Hosseinian, S.H., Dehkordi, G.H.R., Goel, L., Karimyan, P., Prieto-Castrillo, F., Corchado, J.M., Catalão, J.P.: A novel stochastic reserve cost allocation approach of electricity market agents in the restructured power systems. Electr. Power Syst. Res. **152**, 223–236 (2017)
6. Gazafroudi, A.S., Shafie-khah, M., Fitiwi, D.Z., Santos, S.F., Corchado, J.M., Catalão, J.P.: Impact of strategic behaviors of the electricity consumers on power system reliability. In: Sustainable Interdependent Networks II, pp. 193–215. Springer, Cham (2019)
7. Pinto, T., Gazafroudi, A.S., Prieto-Castrillo, F., Santos, G., Silva, F., Corchado, J.M., Vale, Z.: Reserve costs allocation model for energy and reserve market simulation. In: 2017 19th International Conference on Intelligent System Application to Power Systems (ISAP), pp. 1–6. IEEE, September 2017
8. Heydt, G.T.: The next generation of power distribution systems. IEEE Trans. Smart Grid **1**(3), 225–235 (2010)
9. Morente-Molinera, J.A., Kou, G., González-Crespo, R., Corchado, J.M., Herrera-Viedma, E.: Solving multi-criteria group decision making problems under environments with a high number of alternatives using fuzzy ontologies and multi-granular linguistic modelling methods. Knowl. Based Syst. **137**, 54–64 (2017)
10. Li, T., Sun, S., Bolić, M., Corchado, J.M.: Algorithm design for parallel implementation of the SMC-PHD filter. Sig. Process. **119**, 115–127 (2016). https://doi.org/10.1016/j.sigpro.2015.07.013
11. Chamoso, P., Rodríguez, S., de la Prieta, F., Bajo, J.: Classification of retinal vessels using a collaborative agent-based architecture. AI Commun. (Preprint), 1–18 (2018)
12. Chamoso, P., González-Briones, A., Rodríguez, S., Corchado, J.M.: Tendencies of technologies and platforms in smart cities: a state-of-the-art review. Wirel. Commun. Mobile Comput. (2018)
13. Gonzalez-Briones, A., Prieto, J., De La Prieta, F., Herrera-Viedma, E., Corcha-do, J.M.: Energy optimization using a case-based reasoning strategy. Sensors (Basel) **18**(3), 865 (2018). https://doi.org/10.3390/s18030865
14. Gonzalez-Briones, A., Chamoso, P., De La Prieta, F., Demazeau, Y., Corchado, J.M.: Agreement technologies for energy optimization at home. Sensors (Basel) **18**(5), 1633 (2018). https://doi.org/10.3390/s18051633
15. Di Mascio, T., Vittorini, P., Gennari, R., Melonio, A., De La Prieta, F., Alrifai, M.: The learners' user classes in the TERENCE adaptive learning system. In: 2012 IEEE 12th International Conference on Advanced Learning Technologies, pp. 572–576. IEEE, July 2012
16. Tapia, D.I., Alonso, R.S., De Paz, J.F., Zato, C., Prieta, F.D.L.: A telemonitoring system for healthcare using heterogeneous wireless sensor networks. Int. J. Artif. Intell. **6**(S11), 112–128 (2011)
17. de la Prieta, F., Navarro, M., García, J.A., González, R., Rodríguez, S.: Multi-agent system for controlling a cloud computing environment. In: Portuguese Conference on Artificial Intelligence, pp. 13–20. Springer, Heidelberg, September 2013
18. Chamoso, P., Rivas, A., Martín-Limorti, J.J., Rodríguez, S.: A hash based image matching algorithm for social networks. In: Advances in Intelligent Systems and Computing, vol. 619, pp. 183–190. https://doi.org/10.1007/978-3-319-61578-3_18
19. Sittón, I., Rodríguez, S.: Pattern extraction for the design of predictive models in Industry 4.0. In: International Conference on Practical Applications of Agents and Multi-Agent Systems, pp. 258–261 (2017)

20. García, O., Chamoso, P., Prieto, J., Rodríguez, S., De La Prieta, F.: A serious game to reduce consumption in smart buildings. In: Communications in Computer and Information Science, vol. 722, pp. 481–493 (2017). https://doi.org/10.1007/978-3-319-60285-1_41

21. Palomino, C.G., Nunes, C.S., Silveira, R.A., González, S.R., Nakayama, M.K.: Adaptive agent-based environment model to enable the teacher to create an adaptive class. In: Advances in Intelligent Systems and Computing, vol. 617 (2017). https://doi.org/10.1007/978-3-319-60819-8_3

22. Canizes, B., Pinto, T., Soares, J., Vale, Z., Chamoso, P., Santos, D.: Smart city: a GECAD-BISITE energy management case study. In: 15th International Conference on Practical Applications of Agents and Multi-Agent Systems PAAMS 2017, Trends in Cyber-Physical Multi-Agent Systems, vol. 2, pp. 92–100 (2017). https://doi.org/10.1007/978-3-319-61578-3_9

23. Chamoso, P., de La Prieta, F., Eibenstein, A., Santos-Santos, D., Tizio, A., Vittorini, P.: A device supporting the self management of tinnitus. In: Lecture Notes in Computer Science (including subseries Lecture Notes in Artificial Intelligence and Lecture Notes in Bioinformatics), vol. 10209, pp. 399–410 (2017). https://doi.org/10.1007/978-3-319-56154-7_36

24. Román, J.A., Rodríguez, S., de da Prieta, F.: Improving the distribution of services in MAS. In: Communications in Computer and Information Science, vol. 616 (2016). https://doi.org/10.1007/978-3-319-39387-2_4

25. Buciarelli, E., Silvestri, M., González, S.R.: Decision economics. In: Commemoration of the Birth Centennial of Herbert A. Simon 1916–2016 (Nobel Prize in Economics 1978): Distributed Computing and Artificial Intelligence, 13th International Conference. Advances in Intelligent Systems and Computing, vol. 475. Springer (2016)

26. Lima, A.C.E.S., De Castro, L.N., Corchado, J.M.: A polarity analysis framework for Twitter messages. Appl. Math. Comput. **270**, 756–767 (2015). https://doi.org/10.1016/j.amc.2015.08.059

27. Prieto, J., Alonso, A.A., de la Rosa, R., Carrera, A.: Adaptive frame-work for uncertainty analysis in electromagnetic field measurements. Radiation Protection Dosimetry, ncu260 (2014)

28. Chamoso, P., Raveane, W., Parra, V., González, A.: UAVs applied to the counting and monitoring of animals. In: Advances in Intelligent Systems and Computing, vol. 291, pp. 71–80 (2014). https://doi.org/10.1007/978-3-319-07596-9_8

29. Pérez, A., Chamoso, P., Parra, V., Sánchez, A.J.: Ground vehicle detection through aerial images taken by a UAV. In: 2014 17th International Conference on Information Fusion (FUSION) (2014)

30. Choon, Y.W., Mohamad, M.S., Deris, S., Illias, R.M., Chong, C.K., Chai, L.E., Omatu, S., Corchado, J.M.: Differential bees flux balance analysis with OptKnock for in silico microbial strains optimization. PLoS ONE **9**(7) (2014). https://doi.org/10.1371/journal.pone.0102744

31. Li, T., Sun, S., Corchado, J.M., Siyau, M.F.: A particle dyeing approach for track continuity for the SMC-PHD filter. In: FUSION 2014 - 17th International Conference on Information Fusion (2014). https://www.scopus.com/inward/record.uri?eid=2-s2.0-84910637583&partnerID=40&md5=709eb4815eaf544ce01a2c21aa749d8f

32. García Coria, J.A., Castellanos-Garzón, J.A., Corchado, J.M.: Intelligent business processes composition based on multi-agent systems. Expert Syst. Appl. **41**(4 PART 1), 1189–1205 (2014). https://doi.org/10.1016/j.eswa.2013.08.003

33. Heras, S., De la Prieta, F., Julian, V., Rodríguez, S., Botti, V., Bajo, J., Cor-chado, J.M.: Agreement technologies and their use in cloud computing environments. Prog. Artif. Intell. **1** (4), 277–290 (2012)

34. Prieto, J., Mazuelas, S., Bahillo, A., Fernández, P., Lorenzo, R.M., Abril, E.J.: Accurate and robust localization in harsh environments based on V2I communication. In: Vehicular Technologies - Deployment and Applications. INTECH Open Access Publisher (2013)
35. Tapia, D.I., Fraile, J.A., Rodríguez, S., Alonso, R.S., Corchado, J.M.: Integrating hardware agents into an enhanced multi-agent architecture for Ambient Intelligence systems. Inf. Sci. **222**, 47–65 (2013). https://doi.org/10.1016/j.ins.2011.05.002
36. Prieto, J., Mazuelas, S., Bahillo, A., Fernandez, P., Lorenzo, R.M., Abril, E.J.: Adaptive data fusion for wireless localization in harsh environments. IEEE Trans. Signal Process. **60**(4), 1585–1596 (2012)
37. Muñoz, M., Rodríguez, M., Rodríguez, M.E., Rodríguez, S.: Genetic evaluation of the class III dentofacial in rural and urban Spanish population by AI techniques. In: Advances in Intelligent and Soft Computing. AISC, vol. 151 (2012). https://doi.org/10.1007/978-3-642-28765-7_49
38. Costa, Â., Novais, P., Corchado, J.M., Neves, J.: Increased performance and better patient attendance in an hospital with the use of smart agendas. Logic J. IGPL **20**(4), 689–698 (2012). https://doi.org/10.1093/jigpal/jzr021
39. García, E., Rodríguez, S., Martín, B., Zato, C., Pérez, B.: MISIA: middleware infrastructure to simulate intelligent agents. In: Advances in Intelligent and Soft Computing, vol. 91 (2011). https://doi.org/10.1007/978-3-642-19934-9_14
40. Rodríguez, S., De La Prieta, F., Tapia, D.I., Corchado, J.M.: Agents and computer vision for processing stereoscopic images. In: Lecture Notes in Computer Science (including subseries Lecture Notes in Artificial Intelligence and Lecture Notes in Bioinformatics), vol. 6077 (2010). https://doi.org/10.1007/978-3-642-13803-4_12
41. Rodríguez, S., Gil, O., De La Prieta, F., Zato, C., Corchado, J.M., Vega, P., Francisco, M.: People detection and stereoscopic analysis using MAS. In: INES 2010 - 14th International Conference on Intelligent Engineering Systems, Proceedings (2010). https://doi.org/10.1109/INES.2010.5483855
42. Prieto, J., Mazuelas, S., Bahillo, A., Fernández, P., Lorenzo, R.M., Abril, E.J.: On the minimization of different sources of error for an RTT-based indoor localization system without any calibration stage. In: 2010 International Conference on Indoor Positioning and Indoor Navigation (IPIN), pp. 1–6 (2010)
43. Gazafroudi, A.S., Corchado, J.M., Kean, A., Soroudi, A.: Decentralized flexibility management for electric vehicles. IET Renewable Power Generation (2019). http://ietdl.org/t/IBgIPb
44. Gazafroudi, A.S., Soares, J., Ghazvini, M.A.F., Pinto, T., Vale, Z., Corcha-do, J.M.: Stochastic interval-based optimal offering model for residential energy management systems by household owners. Int. J. Electr. Power Energy Syst. **105**, 201–219 (2019)
45. Prieto-Castrillo, F., Shokri Gazafroudi, A., Prieto, J., Corchado, J.M.: An ising spin-based model to explore efficient flexibility in distributed power systems. Complexity (2018)
46. Gazafroudi, A.S., Prieto-Castrillo, F., Pinto, T., Prieto, J., Corchado, J.M., Bajo, J.: Energy flexibility management based on predictive dispatch model of domestic energy management system. Energies **10**(9), 1397 (2017)
47. Najafi, S., Talari, S., Gazafroudi, A.S., Shafie-khah, M., Corchado, J.M., Catalão, J.P.: Decentralized control of DR using a multi-agent method. In: Sustainable Interdependent Networks, pp. 233–249. Springer, Cham (2018)
48. Gazafroudi, A.S., Prieto-Castrillo, F., Pinto, T., Corchado, J.M.: Energy flexibility management in power distribution systems: decentralized approach. In: 2018 International Conference on Smart Energy Systems and Technologies (SEST), pp. 1–6. IEEE, September 2018

49. Ebrahimi, M., Gazafroudi, A.S., Corchado, J.M., Ebrahimi, M.: Energy management of smart home considering residences' satisfaction and PHEV. In: 2018 International Conference on Smart Energy Systems and Technologies (SEST), pp. 1–6. IEEE, September 2018

50. Gazafroudi, A.S., Pinto, T., Prieto-Castrillo, F., Corchado, J.M., Abrishambaf, O., Jozi, A., Vale, Z.: Energy flexibility assessment of a multi agent-based smart home energy system. In: 2017 IEEE 17th International Conference on Ubiquitous Wireless Broadband (ICUWB), pp. 1–7. IEEE, September 2017

51. Navarro-Cáceres, M., Gazafroudi, A.S., Prieto-Castillo, F., Venyagamoorthy, K.G., Corchado, J.M.: Application of artificial immune system to domestic energy management problem. In: 2017 IEEE 17th International Conference on Ubiquitous Wireless Broadband (ICUWB), pp. 1–7. IEEE, September 2017

52. Bajool, R., Shafie-khah, M., Gazafroudi, A.S., Catalão, J.P.: Mitigation of active and reactive demand response mismatches through reactive power control considering static load modeling in distribution grids. In: 2017 IEEE Conference on Control Technology and Applications (CCTA), pp. 1637–1642. IEEE, August 2017

53. Gazafroudi, A.S., Prieto-Castrillo, F., Pinto, T., Jozi, A., Vale, Z.: Economic evaluation of predictive dispatch model in MAS-based smart home. In: International Conference on Practical Applications of Agents and Multi-Agent Systems, pp. 81–91. Springer, Cham, June 2017

54. Nassaj, A., Shahrtash, S.M.: An accelerated preventive agent based scheme for post-disturbance voltage control and loss reduction. IEEE Trans. Power Syst. **33**(44), 4508–4518 (2018)

55. Nassaj, A., Shahrtash, S.M.: A predictive agent-based scheme for post-disturbance voltage control. Int. J. Electr. Power Energy Syst. **98**, 189–198 (2018)

56. Gazafroudi, A.S., De Paz, J.F., Prieto-Castrillo, F., Villarrubia, G., Talari, S., Shafie-khah, M., Catalão, J.P.: A review of multi-agent based energy management systems. In: International Symposium on Ambient Intelligence, pp. 203–209. Springer, Cham, June 2017

57. Gazafroudi, A.S., Pinto, T., Prieto-Castrillo, F., Prieto, J., Corchado, J.M., Jozi, A., Vale, Z., Venayagamoorthy, G.K.: Organization-based multi-agent structure of the smart home electricity system. In: 2017 IEEE Congress on Evolutionary Computation (CEC), pp. 1327–1334. IEEE, June 2017

58. Gazafroudi, A.S., Prieto-Castrillo, F., Corchado, J.M.: Residential energy management using a novel interval optimization method. In: 2017 4th International Conference on Control, Decision and Information Technologies (CoDIT), pp. 0196–0201. IEEE, April 2017

59. González-Briones, A., Chamoso, P., Yoe, H., Corchado, J.M.: Green-VMAS: virtual organization based platform for heating greenhouses using waste energy from power plants. Sensors **18**(3), 861 (2018)

Future of Smart Parking: Automated Valet Parking Using Deep Q-Learning

Nastaran Shoeibi[1](✉) and Niloufar Shoeibi[2](✉)

[1] Babol Noshirvani University of Technology, Babol, Mazandaran, Iran
Nastaran.Shoeibi95@gmail.com
[2] BISITE Research Group, University of Salamanca, Salamanca, Spain
Niloufar.Shoeibi@Usal.es

Abstract. Population growth and increasing the number of vehicles are causing many different economic and environmental problems. One of the crucial ones is finding a parking space. In order to deal with this issue, we can construct new parking lots or optimizing the old ones. In fact, building new parking costs a lot and will destroy nature. Most of the time there is not enough space to build a new one in cities. Thanks to the evolution of IoT, the implementation of smart parking based on IoT is possible. In this paper, We are focusing on an eco-friendly system called Automated Valet Parking which uses hybrid robotic valets in smart parking and helps optimizing parking space usage with Deep Q-Learning which is a reinforcement learning method, In order to achieve high performance. Because reinforcement learning is a goal absorbing algorithm and by well-defining the objective function (The goal), We can achieve optimum policy.

Keywords: IoT · Smart city · Smart transportation · Smart parking · Smart Grid · Robotic Parking System · Automated Valet Parking · Electric vehicle · Reinforcement learning · Deep q-learning

1 Introduction

Technology aims to improve the quality of life and provide the happiness and health of the citizens. The idea of IoT (internet of things) is for describing a smart world where everyday objects and embedded systems have interaction with each other and human beings [1–4]. Therefore, they need to be connected by wire or wireless technologies to Internet-based networks and also need real-time low-cost sensors and actuators [5]. It turns them into smart devices that can compute, communicate and manipulate like a human. as a matter of fact, in the future, incrementally the number of smart devices will be more than their users. Like people's lives, IoT has found its way to cities. Moreover, IoT-based devices are becoming a part of smart cities. They can control and manage environmental issues, traffic congestion and make better public safety and save more energy. There are a lot of researches about it, you can refer to [6]. By the increasing number of vehicles and drivers, Transportation and Parking have become frustrating problem for people and government. It can somehow be solved by building new parking lots and hire valets to guide drivers and also by adding IoT technology to

© Springer Nature Switzerland AG 2020
E. Herrera-Viedma et al. (Eds.): DCAI 2019, AISC 1004, pp. 177–182, 2020.
https://doi.org/10.1007/978-3-030-23946-6_20

Parking Management System we can develop Smart Parking and Robotic Parking System. Integration of Radio Frequency Identification (RFID), Wireless Sensor Network (WSN), Near Field Communication (NFC), Mobile Applications and Smart Grid is being used for implementing a Smart Parking System (SPS). You can read more in [7–18].

2   Related Work

Automated Valet Parking is an Eco-friendly reliable parking system that uses Hybrid and Electric Robots in existing parking lots to park cars into available slots. it will optimize parking lot spaces with the same number of cars to park and minimize the wastage of the fuel and air pollution and it's easy to use. In [19] AVP has been deployed by using k-deque in high-density parking (HDP) into existing parking lots. Ref. [20] presents another IoT architecture using deep-learning for parking spot detection. Stanley bot is the first outdoor robot which uses AVP and is cheaper than construction solutions. In [21], we believe that by changing its learning methodology and using deep q-learning which is a reinforcement learning method, we can make an impressive improvement in AVP's performance.

3   Proposed Method

Reinforcement Learning is one of the most promising directions to build very intelligent robots. It is a way that the agent learns to do a particular task only by itself by doing actions to an environment and see the reward of its action in a feedback loop. Figure 1 shows how an agent interacts with the environment. The entire goal of the agent in this system is to maintain the optimum policy in order to achieve as much reward as possible [22].

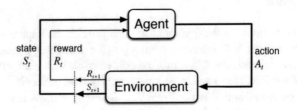

Fig. 1. Markov decision process reinforcement learning [23].

We decided to apply the Q-Learning algorithm to the AVP system because this algorithm is a reinforcement learning algorithm and solve the exact problems as AVP System. Q-Learning is a goal absorbing learning method which performs actions causing the maximum amount of the reward. So gradually it leads to gain the optimum policy. We need to use a deep network because the data which is the situation of the parking is continual, new changes keep happening in the environment and need corresponding reactions. AVP environment is stochastic and every possible action may occur in it. Therefore, the agent needs to learn optimum policy through a process of trial and error. It is clear that the described problem is a Markov Decision Problem (MDP) because it performs actions obeying Markov property [22, 24].

In order to deploy Deep Q-Learning, we need to define States, Actions, and a Reward function. We are proposing a formula to define the performance of the system and shall extract the reward function of the agent from it.

$$Park\ Rate = \frac{Number\ of\ perked\ cars}{Area} \tag{1}$$

$$Performance = \frac{Park\ Rate}{Time \times Moves} \tag{2}$$

Parking rate is a limit to maintain high density it indicates the number of cars parked at a specific area called Parking slot. Park rate, time, number of moves to park a car are variables for defining Performance. on the other words by performance, we apply constraints to the learning of the robot and the aim of the robot is to maximize the performance.

Also, actions are moves the agent can do in order to park the car and states could be the status that whether the path is a dead end, open, or there is an object blocking the way. Having in mind all the positive attributions mentioned above, we believe applying this algorithm can bring remarkable achievements in AVPs.

4 Conclusion and Future Work

It is a fact that the primary costs for constructing AVPs and the costs for taking care of them are a lot, but in long term the benefits are more valuable and undeniable and more users will use this system so we can replace the costs of constructing AVPs after a while. Parking lots will be more densely used and without spending much time and energy users will reach a new level of satisfaction.

One of the things we want to consider in future is less use of power because the robot itself uses power for lifting and parking cars [25].

The other thing is using Multi-agent technologies by adding more than one robot to our AVP system. Refs. [26–50] shows that it can accelerate the learning process and obtaining the optimum policy. By having multi-agent AVP we can speed up the parking process and park more vehicles per hour.

References

1. Casado-Vara, R., Chamoso, P., De la Prieta, F., Prieto, J., Corchado, J.M.: Non-linear adaptive closed-loop control system for improved efficiency in IoT-blockchain management. Inf. Fusion **49**, 227–239 (2019)
2. Casado-Vara, R., Novais, P., Gil, A.B., Prieto, J., Corchado, J.M.: Distributed continuous-time fault estimation control for multiple devices in IoT networks. IEEE Access **7**, 11972–11984 (2019)

3. Casado-Vara, R., Prieto-Castrillo, F., Corchado, J.M.: A game theory approach for cooperative control to improve data quality and false data detection in WSN. Int. J. Robust Nonlinear Control **28**(16), 5087–5102 (2018)
4. Baruque, B., Corchado, E., Mata, A., Corchado, J.M.: A forecasting solution to the oil spill problem based on a hybrid intelligent system. Inf. Sci. **180**(10), 2029–2043 (2010). https://doi.org/10.1016/j.ins.2009.12.032
5. Al-Fuqaha, A., et al.: Internet of Things: a survey on enabling technologies, protocols, and applications. IEEE Commun. Surv. Tutor. **17**(4), 2347–2376 (2015)
6. Chamoso, P., De La Prieta, F.: Swarm-based smart city platform: a traffic application. ADCAIJ Adv. Distrib. Comput. Artif. Intell. J. **4**(2) (2015). ISSN: 2255-2863
7. Mainetti, L., Palano, L., Patrono, L., Stefanizzi, M.L., Vergallo, R.: Integration of RFID and WSN technologies in a Smart Parking System. In: 2014 22nd International Conference on Software, Telecommunications and Computer Networks (SoftCOM) (2014)
8. Gazafroudi, A.S., Soares, J., Ghazvini, M.A.F., Pinto, T., Vale, Z., Corchado, J.M.: Stochastic interval-based optimal offering model for residential energy management systems by household owners. Int. J. Electr. Power Energy Syst. **105**, 201–219 (2019)
9. Prieto-Castrillo, F., Shokri Gazafroudi, A., Prieto, J., Corchado, J.M.: An ising spin-based model to explore efficient flexibility in distributed power systems. Complexity **2018**, 16 (2018). https://doi.org/10.1155/2018/5905932. Article ID 5905932
10. Gazafroudi, A.S., Prieto-Castrillo, F., Pinto, T., Prieto, J., Corchado, J.M., Bajo, J.: Energy flexibility management based on predictive dispatch model of domestic energy management system. Energies **10**(9), 1397 (2017)
11. Gazafroudi, A.S., Shafie-Khah, M., Abedi, M., Hosseinian, S.H., Dehkordi, G.H.R., Goel, L., Karimyan, P., Prieto-Castrillo, F., Corchado, J.M., Catalão, J.P.: A novel stochastic reserve cost allocation approach of electricity market agents in the restructured power systems. Electr. Power Syst. Res. **152**, 223–236 (2017)
12. Gazafroudi, A.S., Shafie-khah, M., Fitiwi, D.Z., Santos, S.F., Corchado, J.M., Catalão, J.P.: Impact of strategic behaviors of the electricity consumers on power system reliability. In: Sustainable Interdependent Networks II, pp. 193–215. Springer, Cham (2019)
13. Gazafroudi, A.S., Prieto-Castrillo, F., Pinto, T., Corchado, J.M.: Energy flexibility management in power distribution systems: decentralized approach. In: 2018 International Conference on Smart Energy Systems and Technologies (SEST), pp. 1–6. IEEE, September 2018
14. Ebrahimi, M., Gazafroudi, A.S., Corchado, J.M., Ebrahimi, M.: Energy management of smart home considering residences' satisfaction and PHEV. In: 2018 International Conference on Smart Energy Systems and Technologies (SEST), pp. 1–6. IEEE, September 2018
15. Pinto, T., Gazafroudi, A.S., Prieto-Castrillo, F., Santos, G., Silva, F., Corchado, J.M., Vale, Z.: Reserve costs allocation model for energy and reserve market simulation. In: 2017 19th International Conference on Intelligent System Application to Power Systems (ISAP), pp. 1–6. IEEE, September 2017
16. Navarro-Cáceres, M., Gazafroudi, A.S., Prieto-Castillo, F., Venyagamoorthy, K.G., Corchado, J.M.: Application of artificial immune system to domestic energy management problem. In: 2017 IEEE 17th International Conference on Ubiquitous Wireless Broadband (ICUWB), pp. 1–7. IEEE, September 2017
17. Bajool, R., Shafie-khah, M., Gazafroudi, A.S., Catalão, J.P.: Mitigation of active and reactive demand response mismatches through reactive power control considering static load modeling in distribution grids. In: 2017 IEEE Conference on Control Technology and Applications (CCTA), pp. 1637–1642. IEEE, August 2017

18. Gazafroudi, A.S., Prieto-Castrillo, F., Corchado, J.M.: Residential energy management using a novel interval optimization method. In: 2017 4th International Conference on Control, Decision and Information Technologies (CoDIT), pp. 0196–0201. IEEE, April 2017
19. Banzhaf, H., Quedenfeld, F., Nienhüser, D., Knoop, S., Zöllner, J.M.: High density valet parking using k-deques in driveways. In: 2017 IEEE Intelligent Vehicles Symposium (IV) (2017)
20. Integration of an automated valet parking service into an internet of things platform. In: 2018 21st International Conference on Intelligent Transportation Systems (ITSC) (2018)
21. Robotics, S.: The first outdoor valet parking robot. https://stanley-robotics.com/
22. Mnih, V., Kavukcuoglu, K., Silver, D., Rusu, A.A., Veness, J., Bellemare, M.G., Graves, A., Riedmiller, M., Fidjeland, A.K., Ostrovski, G., et al.: Human-level control through deep reinforcement learning. Nature **518**, 529 (2015)
23. van Otterlo, M., Wiering, M.: Reinforcement learning and markov decision processes. In: Reinforcement Learning, pp. 3–42. Springer (2012)
24. Gu, S., Lillicrap, T., Sutskever, I., Levine, S.: Continuous deep Q-learning with model-based acceleration. In: International Conference on Machine Learning (2016)
25. Gazafroudi, A.S., Corchado, J.M., Kean, A., Soroudi, A.: Decentralized flexibility management for electric vehicles. IET Renew. Power Gener. (2019). http://ietdl.org/t/IBgIPb
26. Najafi, S., Talari, S., Gazafroudi, A.S., Shafie-khah, M., Corchado, J.M., Catalão, J.P.: Decentralized control of DR using a multi-agent method. In: Sustainable Interdependent Networks, pp. 233–249. Springer, Cham (2018)
27. Gazafroudi, A.S., Pinto, T., Prieto-Castrillo, F., Corchado, J.M., Abrishambaf, O., Jozi, A., Vale, Z.: Energy flexibility assessment of a multi agent-based smart home energy system. In: 2017 IEEE 17th International Conference on Ubiquitous Wireless Broadband (ICUWB), pp. 1–7. IEEE, September 2017
28. Gazafroudi, A.S., Prieto-Castrillo, F., Pinto, T., Jozi, A., Vale, Z.: Economic evaluation of predictive dispatch model in MAS-based smart home. In: International Conference on Practical Applications of Agents and Multi-Agent Systems, pp. 81–91. Springer, Cham, June 2017
29. Gazafroudi, A.S., De Paz, J.F., Prieto-Castrillo, F., Villarrubia, G., Talari, S., Shafie-khah, M., Catalão, J.P.: A review of multi-agent based energy management systems. In: International Symposium on Ambient Intelligence, pp. 203–209. Springer, Cham, June 2017
30. Gazafroudi, A.S., Pinto, T., Prieto-Castrillo, F., Prieto, J., Corchado, J.M., Jozi, A., Vale, Z., Venayagamoorthy, G.K.: Organization-based multi-agent structure of the smart home electricity system. In: 2017 IEEE Congress on Evolutionary Computation (CEC), pp. 1327–1334. IEEE, June 2017
31. González-Briones, A., Chamoso, P., Yoe, H., Corchado, J.M.: GreenVMAS: virtual organization based platform for heating greenhouses using waste energy from power plants. Sensors **18**(3), 861 (2018)
32. Chamoso, P., González-Briones, A., Rivas, A., De La Prieta, F., Corchado, J.M.: Social computing in currency exchange. Knowl. Inf. Syst., 1–21 (2019). https://doi.org/10.1007/s10115-018-1289-4
33. Morente-Molinera, J.A., Kou, G., González-Crespo, R., Corchado, J.M., Herrera-Viedma, E.: Solving multi-criteria group decision making problems under environments with a high number of alternatives using fuzzy ontologies and multi-granular linguistic modelling methods. Knowl. Based Syst. **137**, 54–64 (2017)
34. Li, T., Sun, S., Bolić, M., Corchado, J.M.: Algorithm design for parallel implementation of the SMC-PHD filter. Signal Process. **119**, 115–127 (2016). https://doi.org/10.1016/j.sigpro.2015.07.013

35. Chamoso, P., Rodríguez, S., de la Prieta, F., Bajo, J.: Classification of retinal vessels using a collaborative agent-based architecture. AI Commun. 1–18 (2018). (Preprint)
36. Chamoso, P., González-Briones, A., Rodríguez, S., Corchado, J.M.: Tendencies of technologies and platforms in smart cities: A state-of-the-art review. Wirel. Commun. Mob. Comput. **2018**, 17 (2018). https://doi.org/10.1155/2018/3086854. Article ID 3086854
37. Gonzalez-Briones, A., Prieto, J., De La Prieta, F., Herrera-Viedma, E., Corchado, J.M.: Energy optimization using a case-based reasoning strategy. Sensors (Basel) **18**(3), 865 (2018). https://doi.org/10.3390/s18030865
38. Gonzalez-Briones, A., Chamoso, P., De La Prieta, F., Demazeau, Y., Corchado, J.M.: Agreement technologies for energy optimization at home. Sensors (Basel) **18**(5), 1633 (2018). https://doi.org/10.3390/s18051633
39. Rodríguez, S., De La Prieta, F., Tapia, D.I., Corchado, J.M.: Agents and computer vision for processing stereoscopic images. In: Lecture Notes in Computer Science (including subseries Lecture Notes in Artificial Intelligence and Lecture Notes in Bioinformatics). LNAI, vol. 6077 (2010). https://doi.org/10.1007/978-3-642-13803-4_12
40. Rodríguez, S., Gil, O., De La Prieta, F., Zato, C., Corchado, J.M., Vega, P., Francisco, M.: People detection and stereoscopic analysis using MAS. In: INES 2010 - 14th International Conference on Intelligent Engineering Systems, Proceedings (2010). https://doi.org/10.1109/INES.2010.5483855
41. Rodríguez, S., Tapia, D.I., Sanz, E., Zato, C., De La Prieta, F., Gil, O.: Cloud computing integrated into service-oriented multi-agent architecture. In: IFIP Advances in Information and Communication Technology. AICT, vol. 322 (2010). https://doi.org/10.1007/978-3-642-14341-0_29
42. Tapia, D.I., Corchado, J.M.: An ambient intelligence based multi-agent system for alzheimer health care. Int. J. Ambient. Comput. Intell. **1**(1), 15–26 (2009)
43. Corchado, J.M., Pavón, J., Corchado, E.S., Castillo, L.F.: Development of CBR-BDI agents: a tourist guide application. In: Lecture Notes in Computer Science (including subseries Lecture Notes in Artificial Intelligence and Lecture Notes in Bioinformatics), vol. 3155, pp. 547–559 (2004). https://doi.org/10.1007/978-3-540-28631-8
44. Laza, R., Pavn, R., Corchado, J.M.: A reasoning model for CBR_BDI agents using an adaptable fuzzy inference system. In: Lecture Notes in Computer Science (including subseries Lecture Notes in Artificial Intelligence and Lecture Notes in Bioinformatics), vol. 3040, pp. 96–106. Springer, Heidelberg (2004)
45. Fyfe, C., Corchado Rodríguez, E., González Bedia, M., Corchado Rodríguez, J.M.: Analytical Model for Constructing Deliberative Agents (2002)
46. Rodriguez-Fernandez J., Pinto T., Silva F., Praça I., Vale Z., Corchado J.M.: Reputation computational model to support electricity market players energy contracts negotiation. In: Bajo J., et al. (eds.) Highlights of Practical Applications of Agents, Multi-Agent Systems, and Complexity: The PAAMS Collection. PAAMS 2018. Communications in Computer and Information Science, vol. 887. Springer, Cham (2018)
47. Guimaraes, M., Adamatti, D., Emmendorfer, L.: An agent-based environment for dynamic positioning of the Fogg behavior model threshold line. ADCAIJ Adv. Distrib. Comput. Artif. Intell. J. **7**(1), 67–76 (2018)
48. Omatu, S., Wada, T., Rodríguez, S., Chamoso, P., Corchado, J.M.: Multi-agent technology to perform odor classification. In: ISAmI 2014, pp. 241–252 (2014)
49. Tapia, D.I., Alonso, R.S., García, Ó., Corchado, J.M.: HERA: hardware-embedded reactive agents platform. In: PAAMS (Special Sessions), pp. 249–256 (2011)
50. Souza de Castro, L.F., Vaz Alves, G., Pinz Borges, A.: Using trust degree for agents in order to assign spots in a Smart Parking (2017)

Artificial Intelligence as a Way of Overcoming Visual Disorders: Damages Related to Visual Cortex, Optic Nerves and Eyes

Niloufar Shoeibi[1]([⊠]), Farrokh Karimi[2], and Juan Manuel Corchado[1]

[1] BISITE Research Group, University of Salamanca,
Calle Espejo 2, 37007 Salamanca, Spain
{Niloufar.shoeibi,corchado}@usal.es
[2] Institute for Research in Fundamental Sciences, Artesh Hwy, Tehran, Iran
Farrokh.karimi@sharif.edu

Abstract. Blindness is a disability in which the person got interference with the visual system. In this paper, we studied 18 different diseases. Then categorize the different types of damages to the visual system into three categories which are the brain damages, damages to the eyes and the damages occur to the sensory neurons carrying data. Then suggest possible engineering solutions for each one. Using deep neural networks in order to do the almost same process as the brain does, the developed devices that can be replaced by the vision of the eyes, and also making artificial new connections to cover the attenuation of the sensory neurons carrying data from the eyes to the brain. We are looking forward to overcoming this disability using artificial intelligence apart from medical care. However, we are not denying the importance of medical science but suggesting that new engineering technologies can push up the limits and open new doors to the dead ends. In order to make this dream come true, we need to do lots of studies and will face many more challenges. Some of them are listed in this paper.

Keywords: Blindness · Visual cortex · Brain-computer interface · Artificial neural networks · Deep learning

1 Introduction

Vision disabilities were always considered as a threat to the normal life of all kinds of creatures and gradually lead to blindness. Generally, seeing consists of functioning a lot of parts. The visual system (eyes) are the receptors which gather visual data and does some preprocessing to make the visual data more understandable. The preprocessed data transferred from the eyes to the brain through the sensory neurons called the optic nerves. These kind of neurons are aiming to make a secure connection between the receptors and the processing unit.

The brain has the ability to do object detection by breaking down an image into features like colors, edges, forms, and motion. And simultaneously, or in parallel combine these features to comprehend complicated structures. In [1] how the brain works has been fully explained. The visual cortex is located in the occipital lobe which

© Springer Nature Switzerland AG 2020
E. Herrera-Viedma et al. (Eds.): DCAI 2019, AISC 1004, pp. 183–187, 2020.
https://doi.org/10.1007/978-3-030-23946-6_21

is at the back of the brain. The visual information passes through the thalamus then go to the primary visual cortex. In Sect. 2 we classify the damages related to the visual system and review related work. Then in Sect. 3, we proposed some methods to overcome these damages. Finally, in Sect. 4 we conclude all our studies.

2   Damages to the Visual System and Related Work

In this paper, we study 18 different diseases leading to vision rehabilitation and blindness, to detect which part of the system has got interfered. We categorize them into 3 different groups such as Brain damages, Eyes damages, and Damages related to Optic Nerves, Table 1. Then for each, we reviewed the related work.

Table 1. Diseases categorized based on the damages they cause to the visual system.

Brain	Eyes	Nerves
Head injuries	Retinal detachment	Optic neuritis
Intracranial hemorrhage	Making sense of hypertensive retinopathy	Krabbe disease
Vertebrobasilar circulatory disorders	Cataract	Ito Syndrome
Subarachnoid hemorrhage (SAH)	Macular degeneration	
Brain aneurysm	Strabismus	
Tay-Sachs disease	Eye emergencies	
	Albinism	
	Sarcoidosis	
	Basal cell nevus syndrome	

2.1   Brain Damages

For the brain damages, first of all, it needs to be sure that the damage can be compensated or not. It means that we need to make sure that the damage is only destructing the visual cortex because trying to repair the visual cortex of a dead brain is pointless. In [2] the complexities of experimental closed-loop systems has been studied to have a positive clinical effect on neurological disorders. Moreover, [3] has focused on theoretical impacts and potential benefits of closed-loops. In [4] an accurate functional model for the feedforward, feedback and lateral connections has been observed and a computational model through the thalamus is suggested.

2.2   Eyes Damages

These kind of damages are physical and get interfere with the cornea, pupil, lens, retina, and etc. which cause low vision rehabilitation or blindness. In order to cover dysfunctionality in Low Vision Rehabilitation (LVR) patients, low vision aid devices have been used [5]. There are two kinds of approaches proposed to solve this disability such as Head-Mounted [6] and Heads-Up Display Technologies with totally opposite structures. In Heads-up devices information is shown on a stationary display in the patients' line of sight. On the other hand, Head-Mounted display technology is an electronic visual aid device gathering data from cameras and directly send it to the eyes. However, these devices are not much effective for the patients with serious damages leading to blindness. In [7] the possibility of replacing damaged retina with the artificial silicon retina (ASR) microchip with the objective of safety and efficacy has been studied.

2.3 Damages Related to the Optic Nerves

For the damaged nerves connecting the eyes to the brain, we study the possibility of implanting optical fibers or simply using the electrodes to revoke the neurons in order to reconnect the brain and the eyes. In [8] a method has been proposed for constructing implantable optical fibers to perform with minimum tissue damage and change in light output over time.

3 Proposed Solutions

In this section we suggest using engineering methods to overcome visual disabilities related to brain damages apart from all the medical treatments. Convolutional neural networks (CNN) are a type of artificial neural networks. They are comprised of nodes as neurons and weights as the link between neurons. The output derived from the weighted sum of inputs passed from an activation function. Most new CNNs have several convolutional layers and finally feeds into one or sometimes more fully-connected layers and the final layer of the network performs a multi-class classification, Fig. 1. Generally, CNNs are trained with supervised learning and weights update by backpropagation until network convergence [9].

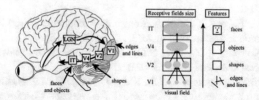

Fig. 1. Hierarchical processing of visual information [10].

It is worth noting that CNNs were directly inspired by the human visual system. CNNs can perform visual tasks at near-human levels. We can map components of CNN to components of the human visual system, Fig. 2.

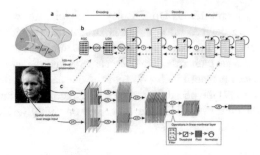

Fig. 2. Hierarchical processing of visual information [10].

4   Conclusion and Future Work

In this paper, we proposed using convolutional neural networks instead of damaged visual layers of the brain to achieve an artificial brain-like processing in order to discover a new sense of seeing for the patients with visual loss disabilities but first we need to face two challenges, finding a convenient convolutional neural network and make a way to connect the network to the brain.

There hasn't been much research about possible solutions for completely vision loss. Totally blind eyes are not functioning at all. Hence, it is crucial to find a suitable alternative. In future, we want to study the possibility of gathering data from other possible receptors like cameras instead of the eyes themselves and finding a way to make an online connection in order to send the data directly to the brain. This proposed solution has its own challenges that are going to be studied.

References

1. Huff, T., Tadi, P.: Neuroanatomy, visual cortex. In: StatPearls [Internet]. StatPearls Publishing (2019)
2. Krook-Magnuson, E., et al.: Neuroelectronics and biooptics: closed-loop technologies in neurological disorders. JAMA Neurol **72**(7), 823–829 (2015)
3. Edward, E.S., Kouzani, A.Z., Tye, S.J.: Towards miniaturized closed-loop optogenetic stimulation devices. J. Neural Eng. **15**, 021002 (2018)
4. George, D., Lavin, A., Guntupalli, J.S., Mely, D., Hay, N., Lazaro-Gredilla, M.: Cortical Microcircuits from a Generative Vision Model. Neurons and Cognition (2018)
5. Chan, T., et al.: Estimates of incidence and prevalence of visual impairment, low vision, and blindness in the United States. JAMA Ophthalmol. **136**(1), 12–19 (2018)
6. Ehrlich, J.R., Ojeda, L.V., Wicker, D., Day, S., Howson, A., Lakshminarayanan, V., Moroi, S.E.: Head-mounted display technology for low-vision rehabilitation and vision enhancement. Am. J. Ophthalmol. **176**, 26–32 (2017)
7. Chow, Alan Y., Chow, Y.Y.C., Packo, Kirk H., et al.: The artificial silicon retina microchip for the treatment of vision loss from retinitis pigmentosa. Arch. Ophthalmol. **122**, 460–469 (2004)

8. Sparta, D.R., Stamatakis, A.M., Phillips, J.L., Hovelsø, N., van Zessen, R., Stuber, G.D.: Construction of implantable optical fibers for long-term optogenetic manipulation of neural circuits. Nat. Protocol. **7**, 12–23 (2011)
9. Kriegeskorte, N.: Deep neural networks: a new framework for modeling biological vision and brain information processing. Annu. Rev. Vis. Sci. **1**, 417–446 (2015)
10. Neurograce: Deep Convolutional Neural Networks as Models of the Visual System: Q & A. Accessed 17 May 2018. https://neurdiness.wordpress.com/

Adjusting the Framework of Multi-agent Systems (MAS) and Internet of Things (IoT) for Smart Power Grids

Amin Nassaj[✉]

Niroo Research Institute (NRI), Tehran, Iran
Aminnasaj@gmail.com

Abstract. In recent years, the artificial intelligence science has introduced the multi-agent system (MAS) which divides a complex problem into some smaller problems, where, the agents in each sub-set must deal with the problem of their area in a cooperative manner. Also, the Internet of things (IoT) technology is one of the novel innovations that has considerable progress and development in intelligence issues in recent years and is referred to as the next technological revolution. Moreover, the wide area monitoring system (WAMS) is a platform for control targets, having global management on power grid contingencies with more effective contributions. In this paper, a uniform structure for adopting the IoT and MAS concepts under the WAMS framework has been described. This novel and attractive structure would be very significant and useful in smart power grid applications.

Keywords: Multi-agent system (MAS) · Internet of Things (IoT) · Wide-Area Monitoring System (WAMS) · Smart grids

1 Introduction

Nowadays, power grids are more equipped with wide area monitoring system (WAMS), which provide dynamic monitoring of power system status. Also, the development of phasor measurement units (PMUs), placed in substations, gives real time information of system condition by synchronized phasors, helps to introducing novel schemes for power grid control purposes. These abilities in dynamically tracking of power system, integration, and processing wide area information can enhance the efficiency of power system, if well-organized indeed [1, 2].

Beside, a decentralized control strategy named as multi-agent system (MAS) has been developed by artificial intelligence researchers. By which a complex problem is divided into some easier sub-problems and each sub-problem is appointed to an agent; while agents placed in an MAS are in negotiation, cooperation and coordination. With the development of the MAS concept, many efforts have been made to employ it in power system studies [1–28].

Moreover, the Internet of things (IoT) technology is one of the novel innovations that has a considerable progress and development in intelligence issues in recent years. The IoT, firstly coined by Kevin Ashton as the title of a presentation in 1999, is a

© Springer Nature Switzerland AG 2020
E. Herrera-Viedma et al. (Eds.): DCAI 2019, AISC 1004, pp. 188–191, 2020.
https://doi.org/10.1007/978-3-030-23946-6_22

technological revolution that is bringing science into a new ubiquitous connectivity, computing, and communication era [29].

This article, a uniform structure for adopting the IoT and MAS concepts under WAMS framework has been described in the following. This novel and attractive structure would be very significant and useful in smart power grid applications.

2 Proposed Approach

Generally, multi-agent systems are systems combined of multiple interacting smart elements, known as agents. The agents can be categorized into two types, including Intelligent Agent (IA) and Reactive Agent (RA) [3, 4]. According to the previous introduced structure in [4], there are phasor measurement units that measure and submit data at a pre-defined frame rate to phasor data concentrators (local PDCs), which have the same role as IAs, with the ability of receiving synchronized data by PMU, processing received data, taking rational decisions (by negotiation/cooperation with other IAs), and dispatching commands to actuators devices. On the other hand, managing the positions and the connectivity of actuators devices (including tap changer transformers, capacitor banks, reactors and switches) are conducted by RAs.

Beside, the overall structure of IoT technology consists of four layers. These are including data collection (with the help of sensors), information transfer (by communication system), information management (using processors and cloud computing approach), and employing this information for making appropriate decisions. It can be seen that the performance of these two domains, i.e. IoT and MAS, have an in-line behaviors and can be integrated in a uniform distributed framework.

In more details, the measurement device related to the IAs (such as PMUs for transmission lines or smart meters for distribution systems), which have the task of collecting data in the power grid, has been defined as Sensors by the technology of IoT that are responsible for monitoring the relevant equipment. Also, the equipment associated with each IAs in the network (such as busbars, transformers, transmission lines and other elements) which are constantly being monitored, play the role of "Things" in the IoT domain. In addition, the decisions of IAs will be applied in power grid by various RAs (such as switches) that act as an "actuator" in the IoT area. Also, both system engage some processing and communication infrastructures to handling the associated system.

Despite this similarity, It seems to be a major difference in these two structures, that is distributed management and decision-making in MAS versus the centralized manner in IoT structure, which need to employing cloud computing. Accordingly, in order to take the advantages of MAS into the IoT structure, this paper proposes a distributed agent-based structure for IoT named as DIoT, has been shown in Fig. 1.

In this structure, depending on the target set for IoT, the proposed structure in Fig. 1 could be implemented for subsets of related things, where, the negotiation mechanism between the superior layers has been predicted. For example, in the Internet of Vehicles (IoV) in smart city, this structure might be employed for private and public vehicles individually, where, every one managing their domain by each subset of conventional IoT, and inter-domain interactions deal with negotiations under DIoT concept.

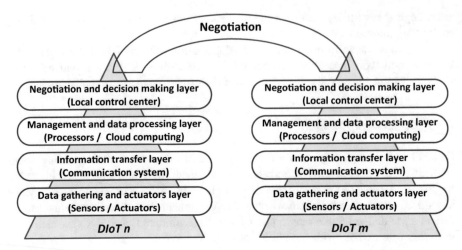

Fig. 1. The structure of proposed DIoT under MAS concept for smart applications.

References

1. Nassaj, A., Shahrtash, S.M.: An accelerated preventive agent based scheme for post-disturbance voltage control and loss reduction. IEEE Trans. Power Syst. **33**(44), 4508–4518 (2018)
2. Nassaj, A., Shahrtash, S.M.: A predictive agent-based scheme for post-disturbance voltage control. Int. J. Electr. Power Energy Syst. **98**, 189–198 (2018)
3. Wooldridge, M.J.: An Introduction to Multi Agent Systems. Wiley, Hoboken (2009)
4. Nassaj, A.: A novel agent-based platform for wide-area monitoring and control in power systems. In: 15th International Conference on Distributed Computing and Artificial Intelligence, Special Sessions, DCAI 2018. Advances in Intelligent Systems and Computing, vol. 801, pp. 385–387. Springer, Cham (2019)
5. Nassaj, A., Shahrtash, S.M.: Confronting with time delays in tap changers for dynamic voltage control by multi agent systems. In: 31st International Power Systems Conference (PSC), Tehran, Iran (2016)
6. Nassaj, A., Shahrtash, S.M.: Prevention of voltage instability by adaptive determination of tap position in OLTCs. In: 25th Iranian Conference on Electrical Engineering (ICEE), Tehran, Iran (2017)
7. Nassaj, A., Shahrtash, S.M.: A dynamic voltage control scheme by employing cooperative game theory. In: 25th Iranian Conference on Electrical Engineering (ICEE), Tehran, Iran (2017)
8. Gazafroudi, A.S., Corchado, J.M., Kean, A., Soroudi, A.: Decentralized flexibility management for electric vehicles. IET Renew. Power Gener. (2019)
9. Gazafroudi, A.S., Soares, J., Ghazvini, M.A.F., et al.: Stochastic interval-based optimal offering model for residential energy management systems by household owners. Int. J. Electr. Power Energy Syst. **105**, 201–219 (2019)
10. Prieto-Castrillo, F., Gazafroudi, A.S., Prieto, J., et al.: An ising spin-based model to explore efficient flexibility in distributed power systems. Complexity (2018)

11. Gazafroudi, A.S., Prieto-Castrillo, F., Pinto, T., et al.: Energy flexibility management based on predictive dispatch model of domestic energy management system. Energies **10**, 1397 (2017)

12. Gazafroudi, A.S., Shafie-Khah, M., Abedi, M., et al.: A novel stochastic reserve cost allocation approach of electricity market agents in the restructured power systems. Electr. Power Syst. Res. **152**, 223–236 (2017)

13. de la Prieta, F., Navarro, M., García, J.A., et al.: Multi-agent system for controlling a cloud computing environment. In: Portuguese Conference on Artificial Intelligence, pp. 13–20. Springer, Heidelberg (2013)

14. García, O., Chamoso, P., Prieto, J., et al.: A serious game to reduce consumption in smart buildings. In: Communications in Computer and Information Science, vol. 722, pp. 481–493 (2017)

15. Román, J.A., Rodríguez, S., de da Prieta, F.: Improving the distribution of services in MAS. In: Communications in Computer and Information Science, vol. 616 (2016)

16. Gazafroudi, A.S., Prieto-Castrillo, F., Pinto, T., et al.: Energy flexibility management in power distribution systems: decentralized approach. In: 2018 International Conference on Smart Energy Systems and Technologies (SEST), pp. 1–6 (2018)

17. Ebrahimi, M., Gazafroudi, A.S., Corchado, J. M., et al.: Energy management of smart home considering residences' satisfaction and PHEV. In: 2018 International Conference on Smart Energy Systems and Technologies (SEST), pp. 1–6 (2018)

18. Pinto, T., Gazafroudi, A.S., Prieto-Castrillo, F., et al.: Reserve costs allocation model for energy and reserve market simulation. In: 19th International Conference on Intelligent System Application to Power Systems (ISAP), pp. 1–6 (2017)

19. Rodríguez, S., Tapia, D.I., Sanz, E., et al.: Cloud computing integrated into service-oriented multi-agent architecture. In: IFIP Advances in Information and Communication Technology, vol. 322 (2010)

20. Tapia, D.I., Corchado, J.M.: An ambient intelligence based multi-agent system for alzheimer health care. Int. J. Ambient. Comput. Intell. **1**, 15–26 (2009)

21. Hernández, E., González, A., Pérez, B., et al.: virtual organization for fintech management. In: International Symposium on Distributed Computing and Artificial Intelligence, pp. 201–210. Springer, Cham (2018)

22. McArthur, S., Davidson, E.M., Catterson, V.M., et al.: Multi-agent systems for power engineering applications-Part I: Concepts, approaches, and technical challenges. IEEE Trans. Power Syst. **22**(4), 1743–1752 (2007)

23. Mehrpooya, M., Mohammadi, M., Ahmadi, E.: Techno-economic-environmental study of hybrid power supply system: a case study in Iran. Sustain. Energy Technol. Assess. **25**, 1–10 (2018)

24. Mohammadi, M., et al.: Optimal planning of renewable energy resource for a residential house considering economic and reliability criteria. Int. J. Electr. Power Energy Syst. **96**, 261–273 (2018)

25. Noorollahi, Y., et al.: Modeling for diversifying electricity supply by maximizing renewable energy use in Ebino city southern Japan. Sustain. Cities Soc. **34**, 371–384 (2017)

26. Noorollahi, Y., Yousefi, H., Mohammadi, M.: Multi-criteria decision support system for wind farm site selection using GIS. Sustain. Energy Technol. Assess. **13**, 38–50 (2016)

27. Mohammadi, M., et al.: Energy hub: from a model to a concept–a review. Renew. Sustain. Energy Rev. **80**, 1512–1527 (2017)

28. Mohammadi, M., et al.: Optimal scheduling of energy hubs in the presence of uncertainty-a review. J. Energy Manag. Technol. **1**(1), 1–17 (2017)

29. Wu, Q., Ding, G., Xu, Y., et al.: Cognitive Internet of Things: a new paradigm beyond connection. IEEE Internet Things J. **1**(2), 129–143 (2014)

Developing a Framework for Distributed and Multi-agent Management of Future Sustainable Energy Systems

Mohammad Mohammadi[✉]

Niroo Research Institute (NRI), Tehran, Iran
Atammy70@gmail.com

Abstract. The development of distributed energy resources (DER), particularly renewable energy sources (RES) and local energy storage systems (ESS), leads to non-hierarchical and distributed structures. On the other hand, the advent of efficient technologies, such as combined heat and power (CHP) production, leads to the integration of energy infrastructure, such as electricity, natural gas, and district heating networks which create a multi-energy system (MES). Energy hub is a promising option for integrated management of MESs. This paper endeavors to outline a realistic model of future sustainable energy systems by integrating RES and taking into account multiple uncertainties. This paper provides a framework for cooperative performance between integrated energy hubs as multi-agent systems (MAS) for distributed management and energy exchange in the presence of RES, controllable loads, and modeling multiple uncertainties. This is a framework that can be considered for future sustainable energy systems, and the main objective of this study is to realize such cooperation of these systems.

Keywords: Multi-Agent System (MAS) · Multi Energy System (MES) · Sustainable energy systems · Energy hub · Smart grid

1 Introduction

In all energy systems, the main objective of customers is to minimize their energy costs while the utilities are concern about load shape, peak load and service quality besides profit [1, 2]. The centralized control method is commonly used for the small-scale systems, and it is not able to handle the large-scale systems [1, 3, 4]. Also in modern electricity markets, the use of centralized management is not feasible for two reasons. First, the privacy of market agents may not be protected, and all participants do not allow the market operator to access all of their information. Second, collecting and processing large amounts of information in a centralized controller is very difficult that needed multi agent-based approach in industries [5].

Due to the proximity of energy hubs and the presence of different components and structures, the probability of different energy hubs with different operators in a real energy system is very high [6]. This could lead to a competitive environment, and a potential for mutual cooperation. In this regard, the decisions of each of the energy hubs can affect the price of local markets and even the decisions of other energy hubs [7].

© Springer Nature Switzerland AG 2020
E. Herrera-Viedma et al. (Eds.): DCAI 2019, AISC 1004, pp. 192–196, 2020.
https://doi.org/10.1007/978-3-030-23946-6_23

However, most studies have focused on the independent planning of energy hubs and does not include cooperation between energy hubs, hence their methods usually provide non-Pareto solutions [8, 9]. Table 1 summarizes a taxonomy of proposed methodologies in the scheduling of SEHs.

Table 1. Comparing the proposed framework with different studies

References	Distributed control	Reliability constraints	Uncertainty				Energy markets	EV	Network model	DR
			Wind	solar	Demand	price				
[10]	✗	✓	✓	✗	✓	✗	✗	✗	✗	✗
[11]	✗	✗	✓	✗	✓	✓	✗	✗	✗	✗
[12]	✓	✗	✗	✗	✗	✗	✗	✗	✗	✓
[13]	✓	✓	✗	✗	✓	✓	✓	✓	✗	✓
[14]	✓	✓	✓	✗	✓	✗	✗	✓	✗	✗
[15]	✗	✗	✗	✓	✗	✗	✗	✓	✓	✓
[16]	✗	✗	✗	✗	✓	✓	✗	✗	✗	✓
[17]	✓	✗	✗	✗	✗	✗	✓	✗	✓	✗
[18]	✓	✗	✗	✗	✗	✗	✓	✗	✗	✗
[19]	✓	✗	✗	✗	✗	✗	✓	✗	✗	✗
[20]	✗	✗	✓	✓	✓	✓	✗	✗	✓	✓
[21]	✗	✗	✗	✗	✓	✓	✓	✗	✗	✗
[22]	✓	✗	✗	✗	✗	✗	✓	✗	✗	✗
proposed	✓	✓	✓	✓	✓	✓	✓	✓	✓	✓

According to the above table, in each of the above research, some of the essential items for the future sustainable energy systems modeling have been disregarded, which leads to unrealistic models. For attaining a realistic model of smart and sustainable energy systems, important challenges such as RES integration, uncertainty modeling, controllable loads, IDR, energy networks, the exchange of power and information, and also the reduction of the computational volume arises in modeling, which should be addressed appropriately [23–32].

2 Proposed Approach

Motivated by the challenges of the above research, this paper endeavors to provide a framework for cooperative performance between integrated energy hubs for distributed management and energy exchange in the presence of RES, controllable loads, and modeling multiple uncertainties.

Energy hubs are assumed to be self-governing, but they tend to participate in a multilateral collaborative environment. In this framework, only the existence of centralized decision-making system is not enough and each component must have the ability to make decisions and manage their tasks as an active agent which can also operate under coordinated and integrated management. It is also assumed that the upstream market price is affected by the cumulative demand of energy hubs. Energy hubs, therefore, are working to realize this impact in a way that they can attain the optimal and favorable energy prices.

The main question, however, is how much this price should be and how energy can be exchanged?

A cooperative framework for integrated energy hubs is used in the form of a bargaining game that can lead to the payoff optimization of energy hubs, the benefits of a cooperative operation to attain a fair agreement for benefit and privacy retention. In this cooperation, each of the energy hubs can participate as producers or consumers. In this regard, energy hub operators can bargain over the price and exchange rate of energy to reach a consensus. Each energy hub has an energy management system that can send and receive control signals to manage the function of its internal components, as well as information signals for the exchange of information with other energy hubs and energy companies. Therefore, it can be said that every energy hub can negotiate energy exchanges with other energy hubs, and at the same time can directly buy electricity and gas from energy utilities. This is a framework that can be considered for future sustainable energy systems, and the main objective of this study has been to actualize such cooperation and modeling of these systems. The solution is to use decentralized and multi-agent management, whose general structure is presented in Fig. 1.

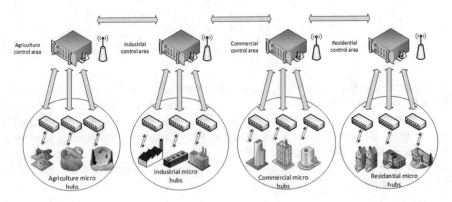

Fig. 1. The structure of proposed multi-agent energy hub management system [33].

References

1. Mehrpooya, M., Mohammadi, M., Ahmadi, E.: Techno-economic-environmental study of hybrid power supply system: a case study in Iran. Sustain. Energ. Technol. Assessments **25**, 1–10 (2018)
2. Mohammadi, M., et al.: Optimal planning of renewable energy resource for a residential house considering economic and reliability criteria. Int. J. Electr. Power Energ. Syst. **96**, 261–273 (2018)
3. Noorollahi, Y., et al.: Modeling for diversifying electricity supply by maximizing renewable energy use in Ebino city southern Japan. Sustain. Cities Soc. **34**, 371–384 (2017)
4. Noorollahi, Y., Yousefi, H., Mohammadi, M.: Multi-criteria decision support system for wind farm site selection using GIS. Sustain. Energ. Technol. Assessments **13**, 38–50 (2016)

5. Mohammadi, M., et al.: Energy hub: from a model to a concept–a review. Renew. Sustain. Energ. Rev. **80**, 1512–1527 (2017)
6. Sepehr, M., et al.: Modeling the electrical energy consumption profile for residential buildings in Iran. Sustain. Cities Soc. **41**, 481–489 (2018)
7. Mohammadi, M., et al.: Optimal scheduling of energy hubs in the presence of uncertainty-a review. J. Energy Manag. Technol. **1**(1), 1–17 (2017)
8. Mohammadi, M., Noorollahi, Y., Mohammadi-Ivatloo, B.: Impacts of energy storage technologies and renewable energy sources on energy hub systems. In: Mohammadi-Ivatloo, B., Jabari, F. (eds.) Operation, Planning, and Analysis of Energy Storage Systems in Smart Energy Hubs, pp. 23–52. Springer, Cham (2018)
9. Mohammadi, M., Noorollahi, Y., Mohammadi-Ivatloo, B.: Demand response participation in renewable energy hubs. In: Mohammadi-Ivatloo, B., Jabari, F. (eds.) Operation, Planning, and Analysis of Energy Storage Systems in Smart Energy Hubs, pp. 129–161. Springer, Cham (2018)
10. Dolatabadi, A., et al.: Optimal stochastic design of wind integrated energy hub. IEEE Trans. Industr. Inf. **13**(5), 2379–2388 (2017)
11. Dolatabadi, A., Jadidbonab, M., Mohammadi-ivatloo, B.: Short-term scheduling strategy for wind-based energy hub: a hybrid stochastic/IGDT approach. IEEE Trans. Sustain. Energy **10**(1), 438–448 (2018)
12. Bahrami, S., Sheikhi, A.: From demand response in smart grid toward integrated demand response in smart energy hub. IEEE Trans. Smart Grid. **7**(2), 650–658 (2016)
13. Bahrami, S., et al.: A decentralized energy management framework for energy hubs in dynamic pricing markets. IEEE Trans. Smart Grid. **9**(6), 6780–6792 (2017)
14. Moeini-Aghtaie, M., et al.: Generalized analytical approach to assess reliability of renewable-based energy hubs. IEEE Trans. Power Syst. **32**(1), 368–377 (2017)
15. Rastegar, M., et al.: A probabilistic energy management scheme for renewable-based residential energy hubs. IEEE Trans. Smart Grid. **8**(5), 2217–2227 (2017)
16. Alipour, M., Zare, K., Abapour, M.: MINLP probabilistic scheduling model for demand response programs integrated energy hubs. IEEE Trans. Industr. Inf. **14**(1), 79–88 (2018)
17. Li, R., et al.: Participation of an energy hub in electricity and heat distribution markets: an MPEC approach. IEEE Trans. Smart Grid., 1–1 (2018)
18. Zhong, W., et al.: ADMM-based distributed auction mechanism for energy hub scheduling in smart buildings. IEEE Access **6**, 45635–45645 (2018)
19. Zhang, X., et al.: Multi-agent bargaining learning for distributed energy hub economic dispatch. IEEE Access **6**, 39564–39573 (2018)
20. Majidi, M., Zare, K.: Integration of smart energy hubs in distribution networks under uncertainties and demand response concept. IEEE Trans. Power Syst. **34**(1), 566–574 (2018)
21. Najafi, A., et al.: A stochastic bilevel model for the energy hub manager problem. IEEE Trans. Smart Grid. **8**(5), 2394–2404 (2017)
22. Fan, S., et al.: Cooperative economic scheduling for multiple energy hubs: a bargaining game theoretic perspective. IEEE Access **6**, 27777–27789 (2018)
23. Nassaj, A., Shahrtash, S.M.: An accelerated preventive agent based scheme for postdisturbance voltage control and loss reduction. IEEE Trans. Power Syst. **33**(4), 4508–4518 (2018)
24. Nassaj, A., Shahrtash, S.M.: Confronting with time delays in tap changers for dynamic voltage control by multi agent systems. In: Proceedings of the 31st International Power System Conference (2016)
25. Nassaj, A., Shahrtash. S.M.: A dynamic voltage control scheme by employing cooperative game theory. In: 2017 Iranian Conference on Electrical Engineering (ICEE). IEEE (2017)
26. Gazafroudi, A.S., et al.: Impact of strategic behaviors of the electricity consumers on power system reliability. In" Sustainable Interdependent Networks II, pp. 193–215. Springer (2019)

27. Nassaj, A.: A novel agent-based platform for wide-area monitoring and control in power systems. In: International Symposium on Distributed Computing and Artificial Intelligence. Springer (2018)

28. Gazafroudi, A.S., et al.: A novel stochastic reserve cost allocation approach of electricity market agents in the restructured power systems. Electric Power Syst. Res. **152**, 223–236 (2017)

29. Nassaj, A., Shahrtash, S.M.: A predictive agent-based scheme for post-disturbance voltage control. Int. J. Electr. Power Energy Syst. **98**, 189–198 (2018)

30. Nassaj, A., Shahrtash, S.M.: Prevention of voltage instability by adaptive determination of tap position in OLTCs. In: 2017 Iranian Conference on Electrical Engineering (ICEE). IEEE (2017)

31. Gazafroudi, A.S., et al.: A review of multi-agent based energy management systems. In: International Symposium on Ambient Intelligence. Springer (2017)

32. Gazafroudi, A.S., et al.: Stochastic interval-based optimal offering model for residential energy management systems by household owners. Int. J. Electr. Power Energy Syst. **105**, 201–219 (2019)

33. Mohammadi, M., et al.: Optimal management of energy hubs and smart energy hubs–a review. Renew. Sustain. Energy Rev. **89**, 33–50 (2018)

Edge Computing:
A Review of Application Scenarios

Inés Sittón-Candanedo$^{(\boxtimes)}$

Bisite Research Group, University of Salamanca,
Calle Espejo 2, 37007 Salamanca, Spain
isittonc@usal.es

Abstract. Edge Computing (EC) emerges as an alternative to process the large volume of data generated by devices connected to the Internet, through the Internet of Things (IoT). This article is an introduction to EC and application scenarios such as: smart cities, Industry 4.0, Smart energy, healthcare, in which EC could allow the optimization of processes that normally run in cloud computing. This optimization consists in the pre-processing the data collected by devices before being sent to a central server, or to the cloud.

Keywords: Edge Computing (EC) · Cloud Computing ·
Internet of Things (IoT)

1 Problem Statement

Up to recently, Cloud Computing was considered one of the best solutions for collecting, processing and analyzing data collected by sensors and Internet of Things devices (IoT) [1–4]. Research points to the challenges of IoT scenarios and sensor networks: security, response times, data quality, latency reduction and bandwidth consumption, among others [5–10]. To address the challenges of IoT and Cloud Computing, this doctoral research aims to propose a solution based on Edge Computing.

The term refers to the fact that the processing of information no longer takes place in a central node (core) but is carried out in the other extreme (edges), displacing the centralization of computing processes from the center of the Internet to its extreme. Likewise, all the information produced by IoT devices is processed in "extreme", releasing computational load on a centralized server, avoiding network traffic overload, and increasing the response time needed for new IoT applications [11]. Edge Computing is characterized by its data microcenters that pre-process and locally store critical data before sending only relevant information to a central data center or cloud storage repository. It is a technology that allows its integration with the Internet of Things by not requiring devices to be constantly connected to the cloud.

1.1 Application Scenarios

By 2019, 45% of the data created on IoT devices is expected to be analyzed by Edge Computing [12]. The benefit of this type of distributed computing approach has been Demonstrated, for example, in facial recognition applications, reducing the response

© Springer Nature Switzerland AG 2020
E. Herrera-Viedma et al. (Eds.): DCAI 2019, AISC 1004, pp. 197–200, 2020.
https://doi.org/10.1007/978-3-030-23946-6_24

time from 900 to 169 ms. [13]; or also in the use of cloudlets (small clouds at the end) used to unload the computing overload of wearable devices, also helping to save between 30% to 40% of energy consumption [14].

Table 1 summarizes the scenarios where the Edge Computing paradigm would offers significant advantages through its implementation.

Table 1. Application scenarios.

Scenarios	References
Industry 4.0, smart manufacturing.	[15–19]
Smart energy	[20–26]
Smart cities	[27–33]
Healthcare	[34, 35]
Agriculture	[36]

2 Reflections

Edge computing and analysis are increasingly being located closer to machines and data sources. According to the case of study, with the advances of industrial systems digitalization, analysis, decision making, and control are physically distributed among the devices: peripheral servers, network, cloud, and connected systems. As a future work, it is suggested to design a hybrid architecture that integrates the advantages of EC and cloud for IoT scenarios, where higher efficiency of the technological infrastructure is required, providing real-time data processing, reduction of data traffic, lower operating costs and higher application performance.

References

1. Casado-Vara, R., Novais, P., Gil, A. B., Prieto, J., Corchado, J.M.: Distributed continuous-time fault estimation control for multiple devices in IoT networks. IEEE Access (2019)
2. De la Prieta, F., Navarro, M., García, J. A., González, R., Rodríguez, S.: Multi-agent system for controlling a cloud computing environment. In: Portuguese Conference on Artificial Intelligence, pp. 13–20. Springer, Heidelberg, September 2013
3. Heras, S., De la Prieta, F., Julian, V., Rodríguez, S., Botti, V., Bajo, J., Corchado, J.M.: Agreement technologies and their use in cloud computing environments. Progress Artif. Intell. **1**(4), 277–290 (2012)
4. Rodríguez-Gonzalez, S., Tapia, D.I., Sanz, E., Zato, C., De La Prieta, F., Gil, O.: Cloud computing integrated into service-oriented multi-agent architecture. In: IFIP Advances in Information and Communication Technology. AICT, vol. 322 (2010)
5. del Rey, Á.M., Batista, F.K., Dios, A.Q.: Malware propagation in wireless sensor networks: global models vs individual-based models. ADCAIJ: Adv. Distrib. Comput. Artif. Intell. J. **6**(3), 5–15 (2017). Salamanca
6. Pinto, A., Costa, R.: Hash-chain-based authentication for IoT. ADCAIJ: Adv. Distrib. Comput. Artif. Intell. J. **5**(4) (2016). Salamanca

7. Becerril, A.A.: The value of our personal data in the big data and the Internet of all Things Era. ADCAIJ: Adv. Distrib. Comput. Artif. Intell. J. **7**(2), 71–80 (2018)
8. Casado-Vara, R., Chamoso, P., De la Prieta, F., Prieto, J., Corchado, J.M.: Non-linear adaptive closed-loop control system for improved efficiency in IoT-blockchain management. Inf. Fusion (2019)
9. Casado-Vara, R., Prieto-Castrillo, F., Corchado, J.M.: A game theory approach for cooperative control to improve data quality and false data detection in WSN. Int. J. Robust Nonlinear Control **28**(16), 5087–5102 (2018)
10. Kethareswaran, V., Sankar Ram, C.: An Indian perspective on the adverse impact of Internet of Things (IoT). ADCAIJ: Adv. Distrib. Comput. Artif. Intell. J. **6**(4), 35–40 (2017)
11. Lopez, P.G., Montresor, A., Epema, D., Datta, A., Higashino, T., Iamnitchi, A., Riviere, E.: *Edge*-centric computing: vision and challenges. SIGCOMM Comput. Commun. Rev. **45**(5), 37–42 (2015)
12. Evans, D.: The internet of things: how the next evolution of the internet is changing everything. CISCO White Paper **1**(2011), 1–11 (2011)
13. Yi, S., Hao, Z., Qin, Z., Li, Q.: Fog computing: platform and applications. In: 2015 Third IEEE Workshop on Hot Topics in Web Systems and Technologies (HotWeb), pp. 73–78 (2015)
14. Ha, K., Chen, Z., Hu, W., Richter, W., Pillai, P., Satyanarayanan, M.: Towards wearable cognitive assistance. In: Proceedings of the 12th Annual International Conference on Mobile Systems, Applications, and Services, pp. 68–81 (2014)
15. Mata, A., Lancho, B.P., Corchado, J.M.: Forest fires prediction by an organization based system. In: PAAMS 2010, pp. 135–144 (2010)
16. Bullon, J., et al.: Manufacturing processes in the textile industry. Expert systems for fabrics production. ADCAIJ: Adv. Distrib. Comput. Artif. Intell. J. **6**(4), 15–23 (2017)
17. Rivas, A., Martín, L., Sittón, I., Chamoso, P., Martín-Limorti, J.J., Prieto, J., González-Briones, A.: Semantic analysis system for industry 4.0. In: International Conference on Knowledge Management in Organizations, pp. 537–548. Springer, Cham, August 2018
18. Sittón-Candanedo, I., Rodríguez-Gonzalez, S.: Pattern extraction for the design of predictive models in industry 4.0. In: International Conference on Practical Applications of Agents and Multi-Agent Systems, pp. 258–261 (2017)
19. Sittón-Candanedo, I., Nieves, E.H., González, S.R., Martín, M.T.S., Briones, A.G.: Machine learning predictive model for industry 4.0. In: International Conference on Knowledge Management in Organizations, pp. 501–510. Springer, Cham, August 2018
20. Fernandes, F., Gomes, L., Morais, H., Silva, M.R., Vale, Z.A., Corchado, J.M.: Dynamic energy management method with demand response interaction applied in an office building. In: PAAMS (Special Sessions) 2016, pp. 69–82 (2016)
21. Gazafroudi, A.S., Corchado, J.M., Kean, A., Soroudi, A.: Decentralized flexibility management for electric vehicles. IET Renew. Power Gener. (2019)
22. Gazafroudi, A.S., Soares, J., Ghazvini, M.A.F., Pinto, T., Vale, Z., Corchado, J.M.: Stochastic interval-based optimal offering model for residential energy management systems by household owners. Int. J. Electr. Power Energy Syst. **105**, 201–219 (2019)
23. Gonzalez-Briones, A., Chamoso, P., De La Prieta, F., Demazeau, Y., Corchado, J.M.: Agreement technologies for energy optimization at home. Sensors (Basel) **18**(5), 1633 (2018)
24. González-Briones, A., Chamoso, P., Yoe, H., Corchado, J.M.: GreenVMAS: virtual organization based platform for heating greenhouses using waste energy from power plants. Sensors **18**(3), 861 (2018)

25. Gonzalez-Briones, A., Prieto, J., De La Prieta, F., Herrera-Viedma, E., Corchado, J.M.: Energy optimization using a case-based reasoning strategy. Sensors (Basel) **18**(3), 865 (2018)
26. Prieto-Castrillo, F., Shokri Gazafroudi, A., Prieto, J., Corchado, J.M.: An Ising spin-based model to explore efficient flexibility in distributed power systems. Complexity (2018)
27. Canizes, B., Pinto, T., Soares, J., Vale, Z., Chamoso, P., Santos, D.: Smart City: a GECAD-BISITE energy management case study. In: 15th International Conference on Practical Applications of Agents and Multi-Agent Systems PAAMS 2017, Trends in Cyber-Physical Multi-Agent Systems, vol. 2, pp. 92–100 (2017)
28. Chamoso, P., González-Briones, A., Rivas, A., De La Prieta, F., Corchado, J.M.: Social computing in currency exchange. Knowl. Inf. Syst. (2019)
29. Chamoso, P., González-Briones, A., Rodríguez, S., Corchado, J.M.: Tendencies of technologies and platforms in smart cities: a state-of-the-art review. Wirel. Commun. Mob. Comput. (2018)
30. Chamoso, P., Rivas, A., Martín-Limorti, J.J., Rodríguez, S.: A hash based image matching algorithm for social networks. In: Advances in Intelligent Systems and Computing, vol. 619, pp. 183–190 (2018)
31. Li, T., Sun, S., Bolić, M., Corchado, J.M.: Algorithm design for parallel implementation of the SMC-PHD filter. Sig. Process. **119**, 115–127 (2016)
32. Morente-Molinera, J.A., Kou, G., González-Crespo, R., Corchado, J.M., Herrera-Viedma, E.: Solving multi-criteria group decision making problems under environments with a high number of alternatives using fuzzy ontologies and multi-granular linguistic modelling methods. Knowl.-Based Syst. **137**, 54–64 (2017)
33. Román, J.Á., Rodríguez, S., Corchado, J.M.: Improving intelligent systems: specialization. In: PAAMS (Workshops) 2014, pp. 378–385 (2014)
34. Chamoso, P., Rodríguez, S., de la Prieta, F., Bajo, J.: Classification of retinal vessels using a collaborative agent-based architecture. AI Commun. (Preprint), 1–18 (2018)
35. Tapia, D.I., Alonso, R.S., De Paz, J.F., Zato, C., Prieta, F.D.L.: A telemonitoring system for healthcare using heterogeneous wireless sensor networks. Int. J. Artif. Intell. **6**(S11), 112–128 (2011)
36. Fuentes, D., Laza, R., Pereira, A.: Intelligent devices in rural wireless networks. DCAIJ: Adv. Distrib. Comput. Artif. Intell. J. **2**(4), 23–30 (2013). Salamanca

A New Approach: Edge Computing and Blockchain for Industry 4.0

Inés Sittón-Candanedo[✉]

Bisite Research Group, University of Salamanca,
Calle Espejo 2, 37007 Salamanca, Spain
isittonc@usal.es

Abstract. The diversity of devices, equipment, sensor networks connected to the Internet, and the paradigm of Industry 4.0; have increased the challenge for the development of architectures, services and intelligent applications capable of processing the data generated constantly. Edge Computing (EC) has emerged as a decentralized alternative and collaborative work for Internet of Things scenarios. However, the security and protection of data has made organizations doubt the effectiveness of EC as well as the model of centralized computing based on the Cloud. This article proposes the integration of Blockchain and Edge Computing technology to obtain a distributed, flexible, scalable and secure data flow through a new reference architecture.

Keywords: Edge Computing · Blockchain · Internet of Things (IoT) · Industry 4.0

1 Problem Statement

Edge Computing emerges as the new technology for the IoT environment where applications require real-time responsiveness with minimal latency, but without interrupting the processing and transfer of the volume of data generated by the devices [1].

For Reyna *et al.,* the Blockchain is a technology that allows the verification and consultation of transactions in a historical way (from the first to the last), structured, open and at any time [2]. The Blockchain system works the storage a set of transactions in each block, which remain linked to each other by a reference to the previous block, building a chain [3, 4].

In the context of Industry 4.0, Edge Computing refers to the technological infrastructure located near the data sources. Its role allows information to be distributed in intermediate nodes avoiding complex data traffic from the edge of the network to the cloud. Nowadays processing and analyzing data in real time is a challenge for the health, telecommunications, manufacturing industries or financial sectors [5]. Therefore, EC gives organizations an advantage by allowing them to preprocess and analyze data in near real time. In the design of an EC platform with IoT for Industry 4.0, data security must be paramount [6]. From this problematic we propose the design of a Global Architecture that integrates Edge Computing and Blockchain for a secure management of the huge amount of data generated by IoT devices, which not only

E. Herrera-Viedma et al. (Eds.): DCAI 2019, AISC 1004, pp. 201–204, 2020.
https://doi.org/10.1007/978-3-030-23946-6_25

collect, but also process the data in real time and are part of this context. Edge Computing contribute to cost reduction of some industrial needs, such as: (i) latency and bandwidth consumption; (ii) maintenance cost, implementation and energy consumption. EC also helps reduce data traveling to the cloud and data vulnerability [1]. The Edge Computing Consortium [7] developed a reference architecture (RA) for Industry 4.0 based on distributed computing and Edge Computing in IoT scenarios. This RA allows the computational load to be seen at nodes in layers closer to the points where the data are being measured. Another reference architecture oriented to industry 4.0 is the one developed by FAR-EDGE that promotes the incorporation of Smart Contracts in its inter-media layer.

2 Reflections

There is a significant incorporation of the Internet of Things (IoT) in scenarios such as: Smart cities [8–11], Smart energy [12–17], smart buildings or home [18]; healthcare [19, 20], rural environments [21], Industry 4.0 [22–24] and its integration with cloud computing [25–27], artificial intelligences techniques [28–31] and multi-agent systems [31–33]. IoT has generated that companies, universities, research centers and governments working on the development of technological solutions, security mechanism [34–37] and architectures that allow them more effective access, visualization and data protection.

Authors propose as future work, the design of a new reference architecture with Edge Computing and Blockchain. Likewise, this new reference architecture will have the main objective to improve the security through the advantages of blockchain technologies, especially in terms of encryption, traceability and data protection from the base layer. This integration also provides reliable, traceable data traffic and privacy control by analyzing sensitive data locally.

References

1. Shi, W., Cao, J., Zhang, Q., Li, Y., Xu, L.: Edge computing: vision and challenges. IEEE Internet Things J. **3**, 637–646 (2016)
2. Reyna, A., Martín, C., Chen, J., Soler, E., Díaz, M.: On blockchain and its integration with IoT. Challenges and opportunities. Futur. Gener. Comput. Syst. **88**, 173–190 (2018)
3. Casado-Vara, R., Chamoso, P., De la Prieta, F., Prieto J., Corchado J.M.: Non-linear adaptive closed-loop control system for improved efficiency in IoT-blockchain management. Information Fusion (2019)
4. Casado-Vara, R., Prieto-Castrillo, F., Corchado, J.M.: A game theory approach for cooperative control to improve data quality and false data detection in WSN. Int. J. Robust Nonlinear Control **28**(16), 5087–5102 (2018)
5. Hernández, E., Sittón, I., Rodríguez, S., Gil, A.B., García, R.J.: An investment recommender multi-agent system in financial technology. In: The 13th International Conference on Soft Computing Models in Industrial and Environmental Applications, pp. 3–10. Springer, Cham, June 2018

6. Hassan, N., Gillani, S., Ahmed, E., Yaqoob, I., Imran, M.: The role of edge computing in Internet of Things. IEEE Commun. Mag. **2018**(99), 1–6 (2018)
7. Edge Computing Consortium, Alliance of Industrial Internet: Edge Computing Reference Architecture 2.0
8. Canizes, B., Pinto, T., Soares, J., Vale, Z., Chamoso, P., Santos, D.: Smart City: a GECAD-BISITE energy management case study. In: 15th International Conference on Practical Applications of Agents and Multi-Agent Systems PAAMS 2017, Trends in Cyber-Physical Multi-Agent Systems, vol. 2, pp. 92–100 (2017)
9. Chamoso, P., González-Briones, A., Rivas, A., De La Prieta, F., Corchado J.M.: Social computing in currency exchange. Knowledge and Information Systems (2019)
10. Chamoso, P., González-Briones, A., Rodríguez, S., Corchado, J.M.: Tendencies of technologies and platforms in smart cities: a state-of-the-art review. Wireless Communications and Mobile Computing (2018)
11. Chamoso, P., Rivas, A., Martín-Limorti, J.J., Rodríguez, S.: A hash based image matching algorithm for social networks. In: Advances in Intelligent Systems and Computing, vol. 619, pp. 183–190 (2018)
12. Gazafroudi, A.S., Corchado, J.M., Kean, A., Soroudi, A.: Decentralized flexibility management for electric vehicles. IET Renewable Power Generation (2019)
13. Gazafroudi, A.S., Soares, J., Ghazvini, M.A.F., Pinto, T., Vale, Z., Corchado, J.M.: Stochastic interval-based optimal offering model for residential energy management systems by household owners. Int. J. Electr. Power Energy Syst. **105**, 201–219 (2019)
14. Gonzalez-Briones, A., Chamoso, P., De La Prieta, F., Demazeau, Y., Corchado, J.M.: Agreement technologies for energy optimization at home. Sensors (Basel) **18**(5), 1633 (2018)
15. González-Briones, A., Chamoso, P., Yoe, H., Corchado, J.M.: GreenVMAS: virtual organization based platform for heating greenhouses using waste energy from power plants. Sensors **18**(3), 861 (2018)
16. Gonzalez-Briones, A., Prieto, J., De La Prieta, F., Herrera-Viedma, E., Corchado, J.M.: Energy optimization using a case-based reasoning strategy. Sensors (Basel), **18**(3), 865 (2018)
17. Prieto-Castrillo, F., Shokri Gazafroudi, A., Prieto, J., Corchado, J.M.: An ising spin-based model to explore efficient flexibility in distributed power systems. Complexity (2018)
18. Casado-Vara, R., Novais, P., Gil, A.B., Prieto, J., Corchado, J.M.: Distributed continuous-time fault estimation control for multiple devices in IoT networks. IEEE Access (2019)
19. Chamoso, P., Rodríguez, S., de la Prieta, F., Bajo, J.: Classification of retinal vessels using a collaborative agent-based architecture. AI Commun. (Preprint), 1–18 (2018)
20. Tapia, D.I., Alonso, R.S., De Paz, J.F., Zato, C., Prieta, F.D.L.: A telemonitoring system for healthcare using heterogeneous wireless sensor networks. Int. J. Artif. Intell. **6**(S11), 112–128 (2011)
21. Fuentes, D., Laza, R., Pereira, A.: Intelligent devices in rural wireless networks. DCAIJ Adv. Distrib. Comput. Artif. Intell. J. **2**(4) (2013)
22. Rivas, A., Martín, L., Sittón, I., Chamoso, P., Martín-Limorti, J.J., Prieto, J., González-Briones, A.: Semantic analysis system for Industry 4.0. In: International Conference on Knowledge Management in Organizations, pp. 537–548. Springer, Cham, August 2018
23. Sittón-Candanedo I., Rodríguez-Gonzalez, S.: Pattern extraction for the design of predictive models in Industry 4.0. In: International Conference on Practical Applications of Agents and Multi-Agent Systems, pp. 258–261 (2017)
24. Sittón-Candanedo, I., Nieves, E.H., González, S.R., Martín, M.T.S., Briones, A.G.: Machine learning predictive model for Industry 4.0. In: International Conference on Knowledge Management in Organizations, pp. 501–510. Springer, Cham, August 2018

25. De la Prieta, F., Navarro, M., García, J.A., González, R., Rodríguez, S.: Multi-agent system for controlling a cloud computing environment. In: Portuguese Conference on Artificial Intelligence, pp. 13–20. Springer, Heidelberg, September 2013

26. Heras, S., De la Prieta, F., Julian, V., Rodríguez, S., Botti, V., Bajo, J., Corchado, J.M.: Agreement technologies and their use in cloud computing environments. Prog. Artif. Intell. **1** (4), 277–290 (2012)

27. Rodríguez-Gonzalez, S., Tapia, D.I., Sanz, E., Zato, C., De La Prieta, F., Gil, O.: Cloud computing integrated into service-oriented multi-agent architecture. In: IFIP Advances in Information and Communication Technology. AICT, vol. 322 (2010)

28. Li, T., Sun, S., Bolić, M., Corchado, J.M.: Algorithm design for parallel implementation of the SMC-PHD filter. Sig. Process. **119**, 115–127 (2016)

29. Morente-Molinera, J.A., Kou, G., González-Crespo, R., Corchado, J.M., Herrera-Viedma, E.: Solving multi-criteria group decision making problems under environments with a high number of alternatives using fuzzy ontologies and multi-granular linguistic modelling methods. Knowl. Based Syst. **137**, 54–64 (2017)

30. Román, J.A., Rodríguez, S., Corchado, J.M.: Improving intelligent systems: specialization. In: International Conference on Practical Applications of Agents and Multi-Agent Systems, pp. 378–385. Springer, Cham, June 2014

31. Tapia, D.I., Fraile, J.A., Rodríguez, S., Alonso, R.S., Corchado, J.M.: Integrating hardware agents into an enhanced multi-agent architecture for Ambient Intelligence systems. Inf. Sci. **222**, 47–65 (2013). https://doi.org/10.1016/j.ins.2011.05.002

32. Román, J.A., Rodríguez, S., de da Prieta, F.: Improving the distribution of services in MAS. In: Communications in Computer and Information Science, vol. 616 (2016). https://doi.org/ 10.1007/978-3-319-39387-2_4

33. Rodríguez, S., De La Prieta, F., Tapia, D.I., Corchado, J.M.: Agents and computer vision for processing stereoscopic images. In: Lecture Notes in Computer Science (including subseries Lecture Notes in Artificial Intelligence and Lecture Notes in Bioinformatics). LNAI, vol. 6077 (2010). https://doi.org/10.1007/978-3-642-13803-4_12

34. del Rey, Á.M., Batista, F.K., Queiruga Dios, A.: Malware propagation in Wireless Sensor Networks: global models vs Individual-based models. ADCAIJ Adv. Distrib. Comput. Artif. Intell. J. **6**(3) (2017)

35. Pinto, A., Costa, R.: Hash-chain-based authentication for IoT. ADCAIJ Adv. Distrib. Comput. Artif. Intell. J. **5**(4) (2016)

36. Becerril, A.A.: The value of our personal data in the Big Data and the Internet of all Things Era. ADCAIJ Adv. Distrib. Comput. Artif. Intell. J. **7**(2), 71–80 (2018)

37. Kethareswaran, V., Sankar Ram, C.: An Indian Perspective on the adverse impact of Internet of Things (IoT). ADCAIJ Adv. Distrib. Comput. Artif. Intell. J. **6**(4), 35–40 (2017)

Internet of Things Platforms Based on Blockchain Technology: A Literature Review

Yeray Mezquita[✉] [iD]

BISITE Research Group, University of Salamanca,
Calle Espejo 2, 37007 Salamanca, Spain
yeraymm@usal.es

Abstract. The implementation of Blockchain Technology (BT) within Internet of Things (IoT) systems has numerous benefits, whereas if it is not correctly adapted to the specific use case, it can present some disadvantages. In the design of IoT Blockchain-based platforms, to identify where the use of BT may present a drawback and search the alternatives that overcome that flaw it is needed to perform a systematic analysis of the platform needs. In order to being capable of face the design of IoT Blockchain-based systems at a later stage, in this work it is performed an analysis of the key characteristics that BT has to take into account when working with it and a literature review of different Blockchain-based IoT platforms.

Keywords: Blockchain · Internet of Things · Literature review

1 Introduction

Our society is evolving into a digitally interconnected world, providing with intelligence the objects we use in our daily lives. This intelligence is provided by sensors and actuators that make the objects capable to read their surroundings and interact with them in an automatic manner [7,14].

The sensors and/or actuators that are associated to a real object, provide data with which a virtual profile of the object is created. Different real objects can be linked via its virtual profiles through Internet, creating a network of interconnected objects that works together optimizing and automating daily life activities. The concept used to define this net of devices that interconnects the real world with the virtual one is Internet of Things (IoT) [3,13,16].

The concept of IoT is very popular in resource optimization problems and highly demanded in today's industry due to its cost savings (Smart Grids, Smart Home, Smart Farming, Smart City) [8,10–12,19]. Although these possibilities make the use of this technology very attractive, they also bring with them some disadvantages [6,15]:

© Springer Nature Switzerland AG 2020
E. Herrera-Viedma et al. (Eds.): DCAI 2019, AISC 1004, pp. 205–208, 2020.
https://doi.org/10.1007/978-3-030-23946-6_26

- Privacy: With devices that monitor virtually all of our activities, companies that have access to the information generated also have access to the routine of our daily lives.
- Security: The continuous exchange of information between the network of devices may cause problems with the integrity of these data.

The previously mentioned problems could be tackled by designing this type of systems with technology capable of ensuring the privacy of the users and the integrity of the data generated. Thanks to the encryption protocol of data end-to-end, Blockchain Technology (BT) have the potential to cover the issues IoT platforms have [1,4].

The blockchain is an incorruptible digital distributed ledger of economic transactions that can be programmed to record not just financial transactions but virtually everything of value. This ledger consists of a peer-to-peer (P2P) network of nodes that keeps the information stored in a redundant way. In an IoT system, BT can be used instead of traditional databases, helping on getting rid of centralized controllers such as banks, accountants and governments [1,5].

The benefits BT brings with itself in IoT systems comes with some downsides:

- Storage capacity and scalability. In BT the size of the data stored in the chain is continually growing. This means that, as time passes and size grows up, the nodes require more resources.
- Consensus. The nodes of the blockchain network need to reach a consensus in order to add the next block of information to the blockchain. The common consensus protocol used in the most famous blockchains of the market (Bitcoin and Ethereum) consumes a great quantity of resources, something that does not match the nature of IoT devices.

The previously mentioned problems can be addressed by changing some characteristics of the BT used in the IoT system. This paper it is going to show a review of how other authors have overcame the problems of combining these two technologies in their approaches. Understanding the current stage of development and usage of this kind of systems.

2 Literature Review

In this sections it is going to be made a review of the works that appeared first while making a search by the keywords "blockchain" and "IoT" in Google Scholar.

In [18] it is shown how BT can be applied to a supply chain in order to strengthen the security of the IoT devices that operates in it. In the case of supply chains, it is used BT to provide transparency and visibility to the transactions made of assets between actors within the supply chain.

In the case of a any kind of IoT system, control and configuration of IoT devices can be made through a blockchain platform [17]. By making use of RSA public key cryptosystems and signatures in the platform, it is possible to avoid attacks such as man-in-the-middle. That's true because messages IoT devices receive are encrypted and signed with the private key of the devices that send them. In this use case, the public key of an IoT device is stored in the blockchain, while the private one is kept inside the device itself.

Making use of smart contracts [2] in addition to the RSA public key cryptosystem, it is possible to avoid also DDOS attacks on the IoT devices. IoT devices are listening just for the blockchain updates, while just devices previously registered can update the blockchain, discarding petitions that don't come from sources that are known by the system.

In order to bring together all the above-mentioned characteristics and requisites of the blockchain-based IoT platforms, a smart home use case scenario is presented in [9]. In that work, for each smart home it is used a local private blockchain, in which registered devices of the smart home read and write data in it. There also exist a IoT device, called Home Miner, that manage the addition or deletion of new devices and the public key cryptosystem. The Home Miner also makes is the one in charge of the interactions between different Smart Homes.

The most used approach in this kind of systems is the one in which the blockchain is not stored inside the IoT devices, rather used as a service from outside the IoT network. This is due to IoT devices being resources-constrained, while the use of a blockchain normally involves the use of many computational resources and bandwidth.

3   Reflections

The use of BT within IoT platforms can be extremely helpful. Helping to avoid man-in-the-middle and DDOS attacks to the system. But the combined integration of those technologies suppose other drawbacks like scalability and energy-resource consumption. Those drawbacks have been faced by the works studied by using BT as a service, using public blockchains that has their own infrastructure like Ethereum or by using local private blockchains.

References

1. Casado-Vara, R., Chamoso, P., De la Prieta, F., Prieto, J., Corchado, J.M.: Nonlinear adaptive closed-loop control system for improved efficiency iniot-blockchain management. Inf. Fusion **49**, 227–239 (2019)
2. Casado-Vara, R., González-Briones, A., Prieto, J., Corchado, J.M.: Smart contract for monitoring and control of logistics activities: pharmaceutical utilities case study. In: The 13th International Conference on Soft Computing Models in Industrial and Environmental Applications, pp. 509–517. Springer (2018)
3. Casado-Vara, R., Novais, P., Gil, A.B., Prieto, J., Corchado, J.M.: Distributed continuous-time fault estimation control for multiple devices in IoT networks. IEEE Access **7**, 11972–11984 (2019)
4. Casado-Vara, R., de la Prieta, F., Prieto, J., Corchado, J.M.: Blockchain framework for IoT data quality via edge computing. In: Proceedings of the 1st Workshop on Blockchain-enabled Networked Sensor Systems, pp. 19–24. ACM (2018)

5. Casado-Vara, R., Prieto, J., Corchado, J.M.: How blockchain could improve fraud detection in power distribution grid. In: The 13th International Conference on Soft Computing Models in Industrial and Environmental Applications, pp. 67–76. Springer (2018)

6. Casado-Vara, R., Prieto-Castrillo, F., Corchado, J.M.: A game theory approach for cooperative control to improve data quality and false data detection in WSN. Int. J. Robust Nonlinear Control. **28**(16), 5087–5102 (2018)

7. Chamoso, P., González-Briones, A., Rodríguez, S., Corchado, J.M.: Tendencies of technologies and platforms in smart cities: a state-of-the-art review. Wirel. Commun. Mob. Comput. **2018**, 1–17 (2018). https://doi.org/10.1155/2018/3086854

8. Chamoso, P., Rodríguez, S., de la Prieta, F., Bajo, J.: Classification ofretinal vessels using a collaborative agent-based architecture. AI Commun., 1–18 (2018). (Preprint)

9. Dorri, A., Kanhere, S.S., Jurdak, R., Gauravaram, P.: Blockchain for IoT security and privacy: the case study of a smart home. In: 2017 IEEE International Conference on Pervasive Computing and Communications Workshops (PerCom Workshops), pp. 618–623. IEEE (2017)

10. Gazafroudi, A.S., Corchado, J.M., Keane, A., Soroudi, A.: Decentralised flexibility management for EVs. IET Renew. Power Gener. (2019). https://doi.org/10.1049/iet-rpg.2018.6023

11. Gazafroudi, A.S., Soares, J., Ghazvini, M.A.F., Pinto, T., Vale, Z., Corchado, J.M.: Stochastic interval-based optimal offering model for residential energy management systems by household owners. Int. J. Electr. Power Energy Syst. **105**, 201–219 (2019). https://doi.org/10.1016/j.ijepes.2018.08.019

12. González-Briones, A., Castellanos-Garzón, J.A., Mezquita Martín,Y., Prieto, J., Corchado, J.M.: A framework for knowledge discovery from wireless sensor networks in rural environments: a crop irrigation systems case study. Wirel. Commun. Mob. Comput. **2018** (2018)

13. González-Briones, A., Chamoso, P., Prieta, F.D.L., Demazeau, Y., Corchado, J.: Agreement technologies for energy optimization at home. Sensors **18**(5), 1633 (2018). https://doi.org/10.3390/s18051633

14. González-Briones, A., Chamoso, P., Yoe, H., Corchado, J.: GreenVMAS: virtual organization based platform for heating greenhouses using waste energy from power plants. Sensors **18**(3), 861 (2018)

15. González-Briones, A., De La Prieta, F., Mohamad, M., Omatu, S., Corchado, J.: Multi-agent systems applications in energy optimization problems: a state-of-the-art review. Energies **11**(8), 1928 (2018)

16. González-Briones, A., Prieto, J., Prieta, F.D.L., Herrera-Viedma, E., Corchado, J.: Energy optimization using a case-based reasoning strategy. Sensors **18**(3), 865 (2018). https://doi.org/10.3390/s18030865

17. Huh, S., Cho, S., Kim, S.: Managing IoT devices using blockchain platform. In: 2017 19th International Conference on Advanced Communication Technology (ICACT), pp. 464–467. IEEE (2017)

18. Kshetri, N.: Can blockchain strengthen the internet of things? IT Prof. **19**(4), 68–72 (2017)

19. Morente-Molinera, J.A., Kou, G., González-Crespo, R., Corchado, J.M., Herrera-Viedma, E.: Solving multi-criteria group decision making problems under environments with a high number of alternatives using fuzzy ontologies and multi-granular linguistic modelling methods. Knowl. Based Syst. **137**, 54–64 (2017)

Deep Learning Techniques for Real Time Computer-Aided Diagnosis in Colorectal Cancer

Alba Nogueira-Rodríguez[1,2(✉)], Hugo López-Fernández[1,2], and Daniel Glez-Peña[1,2]

[1] Department of Computer Science, University of Vigo, ESEI, Campus as Lagoas, 32004 Ourense, Spain
{alnogueira, hlfernandez, dgpena}@uvigo.es
[2] The Biomedical Research Centre (CINBIO), Campus Universitario Lagoas-Marcosende, 36310 Vigo, Spain

Abstract. Colorectal cancer is one of the most common types of cancer. The development of this cancer starts with the presence of polyps or neoplastic lesions in the colon which can evolve to malignant processes. When a polyp is detected during endoscopy, a resection is carried out and a biopsy is done afterwards. Sometimes, resections that have been done are not really necessary, performing an unnecessary procedure over the patient. The PhD project presented here aims develop a real-time colon polyp detection, localization and classification system based on Deep Learning techniques. The creation of this system could help endoscopist in the optical diagnosis of colon lesions, giving an observer-independent aid when making decisions over colorectal cancers.

Keywords: Deep Learning · Colorectal cancer · Computer-aided diagnosis

1 Introduction and Research Statement

Colorectal cancer is one of the most common types of cancer in Spain with 34,331 new patients in 2017 and 15,802 deaths in 2016. The development of this cancer starts with the presence of polyps or neoplastic lesions in the colon which can evolve to malignant processes, so their early detection and subsequent treatment becomes essential. When a polyp is detected during endoscopy, a resection is carried out and a laboratory analysis is done, afterwards (biopsy). Although polyps are removed, new lesions can appear, known as metachronous malignancies, so the follow-up vigilance is necessary. Moreover, sometimes resections that have been done are not really necessary, performing an unnecessary procedure over the patient. The size of the lesions found in most of the individuals is often less than 5 mm, which poses difficulties for visualization, leading to a low detection rate.

In the last decades, different optical technologies were developed to help endoscopists in the determination of the histological diagnosis. However, these technologies require (*i*) magnification endoscopy equipment, (*ii*) the application of stains or pigments and (*iii*) highly experienced endoscopists. To overcome this hurdles, the

© Springer Nature Switzerland AG 2020
E. Herrera-Viedma et al. (Eds.): DCAI 2019, AISC 1004, pp. 209–212, 2020.
https://doi.org/10.1007/978-3-030-23946-6_27

international NICE (Narrow band imaging Colorectal Endoscopic classification) classification was created to be used in conjunction with Narrow Band Imaging (NBI) endoscopes, with or without magnification. These techniques still require highly experienced endoscopists, so there has been a growing interest in the development of computer-aided diagnosis (CADx) systems to predict the histology without the need for polyp resection and biopsies.

CADx systems based in machined learning were developed to classify images of colorectal tumors. For instance, Tamaki et al. developed a computer-aided colorectal tumor classification in NBI endoscopy using local features [1]. The proposed system achieved an accuracy of 96% on a 10-fold cross validation using a real dataset of 908 NBI images and 93% for a separate test dataset. These type of machine learning based systems require an extensive preprocessing of the image datasets in order to extract the features that are used as input by the machine learning algorithms.

In the last years, Deep Learning (DL) has gained a lot of attention in the field of medical image analysis due to its higher performance in image classification tasks when compared to previous state-of-the-art techniques [2]. A recent review collects more than three hundred works on medical image analysis based in DL [3]. Thus, the problem of detecting polyps and predicting its histology can be also addressed using DL models. These models have the advantage that do not require a previous preprocessing of the image datasets and they can be trained using the raw images. The PhD project presented here aims develop a colon polyp detection, localization and classification system based in DL techniques.

2 Related Work

DL has been used to develop successful CADx systems in different fields such as classification of skin cancer [5] and gastrointestinal endoscopy [6]. Also, in the case of colonoscopy, several DL systems based in Convolutional Neuronal Network (CNN) have been developed for the detection and classification of colorectal polyps [4, 7, 8].

Wang et al. developed a DL system based in CNN for the detection of polyps in real time during colonoscopy [7]. The training dataset consisted of 3,634 images with polyps and 1,911 images without polyps (34.46%). The 5,545 images in this dataset were annotated by experienced endoscopists. Then, the algorithm was validated on four independent datasets. The first dataset contained 27,113 images, 5,541 of which included at least one polyp. In this dataset, the algorithm achieved a sensitivity of 94.38% for the detection of polyps and a per-image-specificity of 95.92% and the area under the curve (AUC) of the receiver operating characteristic curve (ROC) was 0.98.

DL requires large amounts of training data to be available in order to build effective predictive models, and this is not always possible, especially in biomedical problems. However, it is possible to apply Transfer Learning (TL), which consists in the usage of pre-trained models and their adaptation via fine-tuning using the data of the problem under consideration so that they can be reused. Zhang et al. [4] applied TL to develop a DL model based on CNN for the automatic detection and classification of colorectal polyps. The dataset contained 1,104 non-polyp images (from both NBI and white-light endoscopy) and 826 polyp images (from NBI endoscopy). The polyp images were classified into 263 hyperplastic polyps and 563 adenomas. The system achieved similar precision when compared with visual inspection by endoscopists (87.3% vbs. 86.4%). Nevertheless, the system achieved higher recall and accuracy rates (87.6% vs. 77.0% and 85.9% vs. 74.3%, respectively). These results prove the feasibility of using TL for these problems. More recently, Urban et al. also developed a DL system for real time identification and localization of polyps testing several TL and non-TL CNNs [8]. The system was trained using 8,641 colonoscopy images containing 4,088 unique polyps and one of the TL CNN was able to identify polyps with a cross-validation accuracy of 96.4% and an AUC of 0.991.

To the best of our knowledge no DL study was carried out tackling detection, localization and classification of colorectal polyps in real time.

3 PhD Research Plan

As stated previously, the PhD project presented here aims develop a colon polyp detection, localization and classification system based in DL techniques. The system, depicted in Fig. 1, will consists of three models: (1) the polyp detection model, which should be able to predict if a given image contains a polyp or not; (2) the polyp localization module, which will take as input an image labeled with "Yes" by the detection model in order to predict the location of the polyp; and (3) the polyp classification model, which should be able to predict the histological diagnosis.

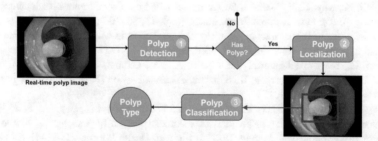

Fig. 1. Overview of the system for colon polyp detection, localization and classification.

To train each model, three differently labeled datasets will be required. This PhD work is framed in a larger project ran within the research group funded by Ministry of Economy, Industry and Competitiveness of Spain. This project is also developing a bank of colorectal polyp images that will be used to generate the training and test datasets for the models developed in this work.

The development of the three models will be done following an iterative process where each iteration serves as feedback for evolving and enhancing the previous models. Initially, a TL approach will be used to develop CNN-based models starting from pre-trained models in non-biomedical data. Well-known models are AlexNet,

VGG16, VGG19 or ResNet, which are available pre-trained with large image collections such as the ImageNet database. This TL approach will enable the creation of models for the PhD work objectives (detection, localization and classification) using the small amounts of data available in the beginning in the bank of colorectal polyp images. The purpose of this initial stage is to obtain a first evaluation and gain insight about the most suitable configurations of the CNN models. Then, CNN-based models will be created from scratch (i.e. without using TL). It is expected that the image bank will contain enough images for enabling the creation of these models. The performance of developed models will be assessed using independent test datasets.

4   Reflections

Early detection of colorectal lesions is key to treat colon cancer successfully. The system proposed in this PhD thesis will be able to perform automatic detection, localization and classification of polyp images in real time. This system would facilitate the work of endoscopists, giving them a second, unbiased opinion. We expect the use of such model will enhance the current polyp detection and identification rates, avoiding unnecessary resections.

References

1. Tamaki, T., Yoshimuta, J., Kawakami, M., Raytchev, B., Kaneda, K., Yoshida, S., Takemura, Y., Onji, K., Miyaki, R., Tanaka, S.: Computer-aided colorectal tumor classification in NBI endoscopy using local features. Med. Image Anal. **17**, 78–100 (2013)
2. Chen, J.H., Asch, S.M.: Machine learning and prediction in Medicine—beyond the peak of inflated expectations. New England J. Med. **376**, 2507–2509 (2017)
3. Litjens, G., Kooi, T., Bejnordi, B.E., Setio, A.A.A., Ciompi, F., Ghafoorian, M., van der Laak, J.A.W.M., van Ginneken, B., Sánchez, C.I.: A survey on DL in medical image analysis. arXiv:1702.05747 [cs] (2017)
4. Zhang, R., Zheng, Y., Mak, T.W.C., Yu, R., Wong, S.H., Lau, J.Y.W., Poon, C.C.Y.: Automatic detection and classification of colorectal polyps by transferring low-level cnn features from nonmedical domain. IEEE J. Biomed. Health Inform. **21**, 41–47 (2017)
5. Esteva, A., Kuprel, B., Novoa, R.A., Ko, J., Swetter, S.M., Blau, H.M., Thrun, S.: Dermatologist-level classification of skin cancer with deep neural networks. Nature **542**, 115–118 (2017)
6. Min, J.K., Kwak, M.S., Cha, J.M.: Overview of deep learning in gastrointestinal endoscopy. Gut Liver. (2019)
7. Wang, P., Xiao, X., Glissen Brown, J.R., Berzin, T.M., Tu, M., Xiong, F., Hu, X., Liu, P., Song, Y., Zhang, D., Yang, X., Li, L., He, J., Yi, X., Liu, J., Liu, X.: Development and validation of a deep-learning algorithm for the detection of polyps during colonoscopy. Nature Biomed. Eng. **2**, 741–748 (2018)
8. Urban, G., Tripathi, P., Alkayali, T., Mittal, M., Jalali, F., Karnes, W., Baldi, P.: DL localizes and identifies polyps in real time with 96% accuracy in screening colonoscopy. Gastroenterology **155**, 1069.e8–1078.e8 (2018)

Trusted Data Transformation
with Blockchain Technology in Open Data

Bruno Tavares[1(✉)], Filipe Figueiredo Correia[1,2], and André Restivo[1,3]

[1] Department of Informatic Engineering, Faculty of Engineering,
University of Porto, Porto, Portugal
{bruno.tavares,filipe.correia,arestivo}@fe.up.pt
[2] INESC TEC, Porto, Portugal
[3] LIACC, Porto, Portugal

Abstract. Trusted open data can be used for auditing, accountability, business development, or as an anti-corruption mechanism. Metadata information can address provenance concerns, and trust issues can somehow be mitigated by digital signatures. Those approaches can trace the data origin, but usually lack information about the transformation process. Creating trust in an open data service through technology can reduce the need for third-party certifications, and creating a distributed consensus mechanism capable of validating all the transformations can guarantee that the datasets are reliable and easy to use. This work aims to leverage blockchain technologies to track open data transformations, allowing consumers to verify the data using a distributed ledger, and providing a mechanism capable of publishing trusted transformed data without relying on third-party certifications. To validate the proposed approach, use cases for data transformation will be used. The consensus protocol must be capable of validating the transformations according to a predefined algorithm, the provider must be capable of publishing verifiable transformed data, and the consumer should be able to check if a dataset originated by a transformation is legit.

Keywords: Data transformation · Open data · Blockchain ·
Distributed ledger

1 Introduction

The ubiquity of technology in our lives is increasing the ability to create large amounts of data. The open data initiative was created to make data free to use, reuse, and redistribute by anyone [6]. Governments and institutions are already making an effort to make data available to the public, in an effort to increase transparency and trust. However, in order to leverage that information, many times, datasets are subject to transformations, resulting in new datasets that can also be made available. The individuals or organizations that wish to use this transformed data must validate the transformations themselves or blindly trust

© Springer Nature Switzerland AG 2020
E. Herrera-Viedma et al. (Eds.): DCAI 2019, AISC 1004, pp. 213–216, 2020.
https://doi.org/10.1007/978-3-030-23946-6_28

the provider. The goal of this work is to propose new research aiming to improve the verification of data transformations in the context of open data. We propose to explore the blockchain technology, to see if it is capable of guaranteeing the provenance of the data and providing an easy way to verify the transformations that it may go through [8, 12].

2   Related Work

Existing works most related to this one are; traditional certifications, open data services, and initiatives that correlate blockchain and open data. As for traditional certifications, the trust is linked back to a trusted Certificate Authority (CA), a digital signature provides cryptographic proof that the certificate has not been modified and has been verified by a trusted third party [15].

Services that offer open data also need to gain a certain amount of trust in order for consumers to use their service. For example, DataGraft is a platform for open data management, it uses a domain-specific language (DSL) to specify transformations that produce linked data graphs. At the time of writing this paper, several of the features identified for future development resolve around transformations, mentioning also better traceability [10].

There are many statements about the applicability of blockchain technology, both positive and negative. There are also research projects that are exploring a way to leverage the blockchain technology to solve open data problems, the SUNFISH project is one example that proposes a cloud federation solution to tackle data governance [9, 11]. The BYTE project refers that data processing has moderate to high priority in the roadmap [5]. The existent regulations must be taken into consideration and systems that use open data such as personal information or locations, must be capable to defend themselves [1–3, 7].

3   Problem

The existence of contextual information is one of the reasons for users to trust or distrust the data available, and for a third party service to be trusted, it must provide some of such information, as provenance. Provenance is the type of metadata that contains information describing the lineage of the data. The integrity of digital records is extremely fragile, it depends on hardware and software that is constantly changing. Most of the time institutions cannot guarantee their availability, usability, and authenticity over time, ending up with untraceable and incomplete data. There are third party institutions known as trusted digital repositories that follow a set of regulatory frameworks that can provide some degree of trust [13]. Another well-known solution to mitigate the lack of trust in open data is the use of certificates. Either way, those approaches follow a traditional path by adding a third party entity to a service that in the end can be compromised. This project aims to leverage the blockchain technology to track open data transformations, this approach will allow consumers to look in the digital ledger in order to verify the data, even if the data is calculated using complex algorithms or is based on a deep tree of calculations. In this work, the correctness and reliability of the original data will not be taken into consideration, the original provider will always be responsible for the data that they make available, and it will always be possible to track the data back to its origin [14].

4   Thesis Statement

When exploring third-party services in open data, concerns regarding trust can arise, primarily when dealing with data transformation. The fundamental research question, for this work, can be stated as:

How can users trust data transformations in a third-party open data service?

The authors claim the following hypothesis:

Blockchain technologies can be used to verify data transformations in third-party open data services. This research should clarify if digital ledgers technologies can improve transparency, and provide trustworthy provenance of data transformations, in an open data context.

The thesis main hypothesis can be decomposed into research goals that can be achieved by addressing the following questions:

- *How can data transformations be tracked in a distributed ledger?*
- *Is it possible to validate data transformations with blockchain technologies?*
- *How can a consumer validate if a dataset has been verified?*

5   Proposal

This project aims to leverage the blockchain technology to track open data transformations, this approach will allow consumers to look into the digital ledger to verify the data, even if the data is calculated using complex algorithms or is based on a deep tree of calculations. In addition, as an incentive mechanism, nodes will generate tokens each time they validate data transformations. In this work, the correctness and reliability of the original data will not be taken into consideration, the original provider will always be responsible for the data that they make available, and it will always be possible to track the data back to the origin. The contributions that this work intends to bring to the community is a blockchain prototype capable of tracking data transformations in open data. If successful this work will present a prototype capable of performing the validation of data transformations of an open data solution. I intend to focus on the case study research methodology, to tackle a real-world problem while simultaneously study the research [4].

To evaluate the proposed solution, it is necessary to first identify several use cases for data transformation, and then verify if the consensus problem can perform the transformations accordingly with a predefined algorithm. The end user should be able to easily check if a dataset originated by a transformation is legit.

6   Conclusions

Data transformations in open data are mostly based on the trust that the consumer has in the service provider. This technology can be used to verify the provenance of the data and what transformation did a specific dataset suffer. Because of the technology intrinsic capabilities, the data trustability can be increased. We feel this project has contemporary significance and can bring value to the academic community, as well as benefits for the open data movement [11].

References

1. Aaronson, S.A.: Data is different: why the world needs a new approach to governing cross-border data flows. CIGI Papers, 197 (2018)
2. Ateniese, G., Magri, B., Venturi, D., Andrade, E.: Redactable blockchain–or–rewriting history in bitcoin and friends. In: Proceedings of IEEE European Symposium on Security and Privacy (EuroS&P), pp. 111–126 (2017)
3. Biswas, K., Muthukkumarasam, V.: Securing smart cities using blockchain technology. In: 2016 IEEE 18th International Conference on High Performance Computing and Communications, pp. 5–6 (2016)
4. Easterbrook, S., Singer, J., Storey, M.-A., Damian, D.: Selecting empirical methods for software engineering research. In: Guide to Advanced Empirical Software Engineering, pp. 285–311 (2008)
5. Grumbach, S., Faravelon, A., Cuquet, M., Fensel, A.: BYTE policy and research roadmap. byte-project.eu (2014)
6. Hannemann, J., Kett, J.: Linked data for libraries meeting: 149. Information technology, cataloguing, classification and indexing with knowledge management. In: World Library and Information Congress: 76 th IFLA General Conference and Assembly (2010)
7. Smith, J., Tennison, J., Wells, P., Fawcett, J., Harrison, S.: Applying blockchain technology in global data infrastructure. Open Data Inst. (2016)
8. Lemieux, V., Lemieux, V.L.: Blockchain and distributed ledgers as trusted record-keeping systems: an archival theoretic evaluation framework. In: Future Technologies Conference (FTC), November (2017)
9. Margheri, A., Schiavo, F.P., Vladimiro, S., Nicoletti, L.: FaaS: federation-as-a-service (Technical report), SUNFISH project (EU-Horizon2020), pp. 1–56. arxiv.org (2014)
10. Roman, D., Nikolov, N., Putlier, A., et al.: DataGraft: one-stop-shop for open data management. Semant. Web 9(4), 393–411 (2018)
11. Singh, S.: A blockchain-based decentralized application for user-driven contribution to Open Government Data. Researchgate.Net, November (2018)
12. Tavares, B., Correia, F.F., Restivo, A., Faria, J.P., Aguiar, A.: A survey of blockchain frameworks and applications. In: The 14th International Conference on Information Assurance and Security (IAS 2018), pp. 1–10 (2018)
13. Thurston Catherine, A.: Trustworthy records and open data. J. Community Inform. 8(2), 2019 (2012)
14. Toussaint, F., Dkrz, S., Atkinson, M., Nerc, K.: Data provenance and tracing for environmental sciences: system design. Envriplus.Eu (2018)
15. Wang, Z., Lin, J., Cai, Q., Wang, Q., Jing, J., Zha, D.: Blockchain-based certificate transparency and revocation transparency. In: Financial Cryptography Workshops (2018)

Electronic Textiles for Intelligent Prevention of Occupational Hazards

Sergio Márquez Sánchez(✉)

Bisite Research Group, University of Salamanca,
Calle Espejo 2, 37007 Salamanca, Spain
smarquez@usal.es
http://bisite.usal.es/es

Abstract. Wearable technologies are taking on increased importance in workspaces, because with then we can monitor the state of people and their environment and react appropriately according to the parameters they capture at real time. They are considered ideal for personal protection equipment doing risk checking and providing support and information to the worker for decision making. We identify solutions made available by industry 4.0 to prevent hazards with a wireless model consists of the design of different innovative PPE (Personal protective equipment) that incorporates intelligent tools and fabrics capable of reacting in real time to a risk situation. Seeking to improve the health and safety of work sectors where there is a high risk of an accident.

Keywords: Smart Textile · Condition monitoring ·
Intelligent environment e-Health · Context modelling

1 Introduction

According to different studies and statistics every day 2 people die as a result of an accident at work, 11 suffer a serious accident, 1391 a minor accident and 1,954 an accident that does not cause sick leave. If we look for the origin of this high accident rate and the upward trend recorded by studies, we can conclude that after a long period of crisis with budget adjustments, precarious work and temporary have favored the hiring of personnel with a lower level of preparation and the increase in work-related accidents [2,3]. The reduction in costs in all areas has resulted in a lack of training and information on occupational risks, which is essential to guarantee their safety and health at work. The types of unexpected risks that may emerge can be: technical, environmental, chemical, physical, intrusive, etc. For example, some of the risks that might appear would be: Installations and/or equipment in poor condition, malfunction, use of inadequate PPE, slips and trips, interior lighting, chemical agents, thermal overload, falling objects, variation in the anthropometric constants of the worker, shocks and blows [1,6–8].

© Springer Nature Switzerland AG 2020
E. Herrera-Viedma et al. (Eds.): DCAI 2019, AISC 1004, pp. 217–220, 2020.
https://doi.org/10.1007/978-3-030-23946-6_29

In this way, this proposal aims to provide an integral solution to improve safety and health in the workplace, providing an innovative tool in the form of an PPE with the capacity to respond in real time to the risks present in the workplace, developing a holistic system, personalized and adapted both to the environment, as well as to the personal circumstances and capabilities of each worker. Nowadays, the available technology allows to approach these problems from a similar approach, but abandoning the traditional training actions and the obsolete PPE, disconnected from the real environment where the working day takes place and lacking in personalization and adaptation to the particular needs of each worker [3–5].

Here, a novel work based on "Smart PPE" that incorporates a network of sensors for monitoring the status. Currently, the rate of accidents among workers is high in all types of industries. Most of those accidents are caused by the lack of prevention measures, poor safety training and obsolete safety systems which do no adapt technologically to the needs of today's workplace. In this work is presented a intelligent system providing workers the tools they need to prevent a large percentage of accidents and create safe working environments. The devices incorporate a suitable technological framework through the use of intelligent fabrics, microelectronics and advanced sensors [9–12].

2   Conclusion

This work has presented a novel system based on the creation of a PRL devices, there's currently no such thing as this system. It combines wireless communication, Smart Textiles and electronic devices. Moreover, the data collected by the system can be visualized by the user on a tablet or a mobile phone. It has been proven that the proposed system significantly improves the safety conditions of operators and also our system allows the data collected to be trained for the early detection of employers who may mean a risk [13–21].

Our system proposes solutions that cover the whole range of action of the strategy of occupational risk prevention, providing advanced means of training to improve safety and health conditions (e.g. incorporation of motion capture technology to perform postural exercises), health improvement (e.g. with sensory suits to learn the best postural practices) and risk prevention (e.g. sensors and alerts both for the worker and for the people around him). Thanks to our knowledge of technology in the fields of circuit printing, printed electronic, stamping of circuits, flexible PCBs or machine learning in a future work we are going to include and improved the quality and characteristics of the technology that we implement in the natural PPE [22–30].

Acknowledgments. This work was supported by the Spanish Junta de Castilla y León, Consejería de empleo. Project: UPPER, aUgmented reality and smart personal protective equipment (PPE) for intelligent pRevention of occupational hazards and accessibility INVESTUN/18/SA/0001.

References

1. Dogan, O., Akcamete, A.: Detecting falls-from-height with wearable sensors and reducing consequences of occupational fall accidents leveraging IoT. In: Advances in Informatics and Computing in Civil and Construction Engineering, pp. 207–214. Springer, Cham (2019)
2. European Agency for Safety and Health at Work (EUOSHA): Priorities for occupational safety and health research in Europe: 2013–2020. Bilbao: EU-OSHA (2013)
3. Podgorski, D., Majchrzycka, K., Dabrowska, A., Gralewicz, G., Okrasa, M.: Towards a conceptual framework of OSH risk management in smart working environments based on smart PPE, ambient intelligence and the Internet of Things technologies. Int. J. Occup. Saf. Ergon. **23**(1), 1–20 (2017)
4. PEROSH: Sustainable workplaces of the future – European research challenges for occupational safety and health. Brussels: PEROSH (Partnership for European Research on Occupational Safety and Health) (2012)
5. Mondal, S.: Phase change materials for smart textiles – an overview. Appl. Therm. Eng. **28**, 1536–1550 (2008)
6. Rocha, J.G., Goncalves, L.M., Rocha, P.F., et al.: Energy harvesting from piezoelectric materials fully integrated in footwear. IEEE Trans. Ind. Electron. **57**, 813–819 (2010)
7. Tao, X.M.: Introduction. In: Tao, X.M. (ed.) Wearable Electronics and Photonics, pp. 1–12. Woodhead in association with The Textile Institute, Cambridge (2005)
8. Scott, R.A. (ed.): Textiles for Protection. Elsevier, Amsterdam (2005)
9. Stephanidis, C.: Human factors in ambient intelligence environments. In: Handbook of Human Factors and Ergonomics, pp. 1354–1373 (2012)
10. Schneegass, S., Voit, A.: GestureSleeve: using touch sensitive fabrics for gestural input on the forearm for controlling smartwatches. In: Proceedings of the 2016 ACM International Symposium on Wearable Computers, ISWC 2016, pp. 108–115. ACM, New York (2016)
11. Gopalsamy, C., Park, S., Rajamanickam, R., Jayaraman, S.: The wearable motherboard: the first generation of adaptive and responsive textile structures (ARTS) for medical applications. Virtual Real. **4**(3), 152–168 (1999)
12. Amft, O., Habetha, J.: Smart medical textiles for monitoring patients with heart conditions. In: Langenhove, L.v. (ed.) Book Chapter in: Smart Textiles for Medicine and Healthcare, pp. 275–297. Woodhead Publishing Ltd., Cambridge, February 2007. ISBN 1 84569 027 3
13. Schneegass, S., Hassib, M., Zhou, B., Cheng, J., Seoane, F., Amft, O., Lukowicz, P., Schmidt, A.: SimpleSkin: towards multipurpose smart garments. In: Adjunct Proceedings of the 2015 ACM International Joint Conference on Pervasive and Ubiquitous Computing and Proceedings of the 2015 ACM International Symposium on Wearable Computers. UbiComp/ISWC 2015 Adjunct, pp. 241–244. ACM, New York (2015)
14. Choi, S., Jiang, Z.: A wearable cardiorespiratory sensor system for analyzing the sleep condition. Expert Syst. Appl. **35**(12), 317–329 (2008)
15. Chen, D., Lawo, M.: Smart textiles and smart personnel protective equipment. In: Smart Textiles, pp. 333–357. Springer, Cham (2017)
16. González-Briones, A., Chamoso, P., Yoe, H., Corchado, J.M.: GreenVMAS: virtual organization based platform for heating greenhouses using waste energy from power plants. Sensors **18**(3), 861 (2018)

17. González-Briones, A., Rivas, A., Chamoso, P., Casado-Vara, R., Corchado, J.M.: Case-based reasoning and agent based job offer recommender system. In: The 13th International Conference on Soft Computing Models in Industrial and Environmental Applications, pp. 21–33. Springer, Cham, June 2018

18. Casado-Vara, R., Prieto-Castrillo, F., Corchado, J.M.: A game theory approach for cooperative control to improve data quality and false data detection in WSN. Int. J. Robust Nonlinear Control **28**(16), 5087–5102 (2018)

19. Rodríguez, S., Palomino, C.G., Chamoso, P., Silveira, R.A., Corchado, J.M.: How to create an adaptive learning environment by means of virtual organizations. In: International Workshop on Learning Technology for Education in Cloud, pp. 199–212. Springer, Cham, August 2018

20. Jia, N.: Detecting human falls with a 3-axis digital accelerometer. In: Volume 43, Number 3, 2009 A Forum for the Exchange of Circuits, Systems, and Software for Real-world Signal Processing, p. 3, July 2009

21. Casado-Vara, R., Novais, P., Gil, A.B., Prieto, J., Corchado, J.M.: Distributed continuous-time fault estimation control for multiple devices in IoT networks. IEEE Access **7**, 11972–11984 (2019)

22. Chamoso, P., González-Briones, A., Rivas, A., De La Prieta, F., Corchado, J.M.: Social computing in currency exchange. Knowl. Inf. Syst., 1–21 (2019)

23. Casado-Vara, R., Prieto-Castrillo, F., Corchado, J.M.: A game theory approach for cooperative control to improve data quality and false data detection in WSN. Int. J. Robust Nonlinear Control. **28**(16), 5087–5102 (2018)

24. Morente-Molinera, J.A., Kou, G., González-Crespo, R., Corchado, J.M., Herrera-Viedma, E.: Solving multi-criteria group decision making problems under environments with a high number of alternatives using fuzzy ontologies and multi-granular linguistic modelling methods. Knowl. Based Syst. **137**, 54–64 (2017)

25. Li, T., Sun, S., Bolić, M., Corchado, J.M.: Algorithm design for parallel implementation of the SMC-PHD filter. Signal Process. **119**, 115–127 (2016). https://doi.org/10.1016/j.sigpro.2015.07.013

26. Chamoso, P., Rodríguez, S., de la Prieta, F., Bajo, J.: Classification of retinal vessels using a collaborative agent-based architecture. AI Commun., 1–18 (2018). (Preprint)

27. Chamoso, P., Rivas, A., Martín-Limorti, J.J., Rodríguez, S.: A hash based image matching algorithm for social networks. In: Advances in Intelligent Systems and Computing, vol. 619, pp. 183–190 (2018). https://doi.org/10.1007/978-3-319-61578-3_18

28. Sittón, I., Rodríguez, S.: Pattern extraction for the design of predictive models in industry 4.0. In: International Conference on Practical Applications of Agents and Multi-Agent Systems, pp. 258–261 (2017)

29. García, O., Chamoso, P., Prieto, J., Rodríguez, S., De La Prieta, F.: A serious game to reduce consumption in smart buildings. In: Communications in Computer and Information Science, vol. 722, pp. 481–493 (2017). https://doi.org/10.1007/978-3-319-60285-1_41

30. Palomino, C.G., Nunes, C.S., Silveira, R.A., González, S.R., Nakayama, M.K.: Adaptive agent-based environment model to enable the teacher to create an adaptive class. In: Advances in Intelligent Systems and Computing, vol. 617 (2017). https://doi.org/10.1007/978-3-319-60819-8_3

Author Index

© Springer Nature Switzerland AG 2020
E. Herrera-Viedma et al. (Eds.): DCAI 2019, AISC 1004, pp. 221–222, 2020.
https://doi.org/10.1007/978-3-030-23946-6

Printed in the United States
By Bookmasters